醉美风尚

国际软装设计流行趋势

Goes Fascinating and Aesthetic the Global Upholstering

欧朋文化 策划　黄滢 马勇 主编

華中科技大學出版社
http://www.hustp.com
中国·武汉

CONTENTS 目录

A

色彩流行趋势

Trend of Color Popularity

B

材料流行趋势

Trend of Material Popularity

E

灯具流行趋势

Trend of Lighting Popularity

C

装饰流行趋势

Trend of Accessory Popularity

D

家具流行趋势

Trend of Furniture Popularity

F

风格流行趋势

Trend of Style Popularity

色彩流行趋势

Trend of Color Popularity

色彩是精神世界的产物，一开始，它被设计师运用于时尚触觉最敏锐、最活跃的服装领域，从而带动起最新的流行趋势与流行色概念，并自然而然地影响着纺织、服装等行业的整体风向，从服装到家纺，从家纺到家居，与室内相关的产品也随之被整体的色彩趋势所渗透与影响。

对于家居而言，流行色讲的是一个“调”字，如果说“调”没有把握好，那么设计便会“满盘皆输”。每年流行色趋势的发布，就如同讲述一个颜色背后令人向往的故事，在众多色彩之中，异军突起的几种颜色及与其进行搭配的那组颜色的背后都隐藏着一个令人期待的故事，当承载着这个故事与文化的色彩被20%~30%的消费者接纳时，它才可被冠以“流行”之名。

色彩不仅是无形的，更是感性的，它是软装设计的精髓与灵魂，通过色相、纯度、色调、对比等手段表达人们的情感和想象，影响人们的心理和生理感受，甚至影响人们对事物的看法。对于室内家居而言，色彩的运用更显其重要性与挑战性，作为人们日常生活的重要场所，家居空间呈现了一个多维空间的立体范畴，区别于传统面料的平面概念，不仅需要考虑产品之间的配色，更需要注重把控整体的空间风格与情景色调。对于家居，我们先要把控好风格，在风格之下再选定具体的情景色调，以风格统一并带动空间色调，注重呈现空间的多色组合。

随着城市生活节奏的加快、人们生活方式的日新月异，如今优秀的家居色彩不仅要具备功能性，更要符合未来人群的生活方式和潜在的审美需求。在2015年国际家居设计流行趋势发布会上，中国流行色协会设计总监张昕曈讲到，2015年国际流行色的五大主题，分别是朦胧清新色、过滤色、充满生机活力的绿色、富有神秘魔幻的色彩、不规则富于变化的色彩。2015年流行色中国色谱则由蓝色主题、绿色主题、粉色主题、风情主题组成。在色彩应用方面，设计强调可灵活运用彩度相似性、色位相似性等，同时要关注明度、流行色、墙面颜色变化。

而擅长透过布艺与色彩搭配赋予空间全新生命力的荷兰色彩大师Lousmijn van den Akker更依据其长期对织物面料的透彻研究，归纳出2016年色彩趋势的六大主题，分别是“沙漠与影”、“美丽的威尼斯”、“美丽的意大利”、“部落风情”、“茶道文化”、“享乐天堂”。

Lousmijn预言，2016年将吹起白色复古风，“沙漠与影”的主题来自于“如何居住在沙漠中

的颜色，深紫色，彩度高，也像纽约的布鲁克林区，以及南非的动物元素，运用黑与白，强烈对比色，也表现在做工精美的刺绣上，让室内空间就像是非洲原住民的舞台，用生动活泼的舞蹈跳出来。

“茶道文化”强调沉稳、厚重，其中的靛蓝色将是未来的趋势，墨绿色也是具有东方禅风的元素，叶子、日本的手做和服都是其灵感来源，追求安静、宁静的禅风深沉空间。而“享乐天堂”则丢掉保守的颜色，带一点金属、反光性质的面料，鲜明、夸张，给人以银光、珠光的感觉，游泳池的颜色也是其中的重要元素。

当然，谈及色彩，不得不说时尚界，因为不管是产品外观的设计，还是室内家装的色彩选择，甚至整个设计界，皆受到时尚界的色彩趋势的影响。以下是根据时尚界、国际权威色彩机构潘通（PANTONE）以及全球性的时尚潮流

的概念”，沙漠里白天有浅的白色、金色，傍晚是咖啡色，面料质地则像沙漠中被风侵蚀过的岩石，充满天然的粗糙触感，就像麻或天然皮，纹理凹凸不平。家具上则以具有穿透性、通透性的家具为代表，另外，古希腊陶罐上的图纹也是灵感来源，将橘色、咖啡色等带入“沙漠与影”的主题中。

“美丽的威尼斯”是非常粉嫩的色系，就像乘着如马卡龙般颜色的贡多拉船行经美丽的古城。此主题与现代感的沙漠主题不同，围绕着“浪漫”两字展开，花卉和蕾丝是构筑浪漫的两个最重要元素，同时丝绒也扮演相当重要的角色。除了清晨的运河，威尼斯的盛大宏伟表现在嘉年华里，而具上图纹概念的运用，很能增添空间中的趣味。

“美丽的意大利”强调颜色鲜明、活泼有力，灵感源自意大利的小渔村，居民会为房子漆上自己喜欢的颜色，房子依山而建，像冰淇淋一样色彩丰富，又像是小渔村五彩缤纷的新鲜水果，很快乐、轻松，很适合多姿多彩的面料和设计。

Lousmijn 表示，明亮的颜色将会回到室内设计行业。“部落风情”有来自非洲部落强烈、明亮

趋势预测机构和专业资讯供应商 Fashion Snoops 发布的流行色趋势总结出的 2015—2016 年的家居流行色彩。

红色（Poppy）：红色是春夏必备的一种颜色，如同罂粟花一样耀眼夺目。红色的空间是对传统文化的继承和发扬，也就是溯美学。

铁锈：仿佛来自沙漠的中性色调，铁锈色将成为最前沿的时尚色彩。这种颜色很容易让人产生一种异域风情的感觉，粗犷奔放，同时还拥有热情。这样的色彩即便是用在家居环境之中，也能让人感受到一种扑面而来的热浪。同时难得的是，它还能制造出一种时光的感觉，长久而令人着迷。

吉士：柔和的黄色带来奶油一样轻柔的质感。这种颜色其实是近几年渐渐流行的，浓浓的奶油般的气息，柔和的色调让人感受到了一股清新的魅力。在家居中，这种颜色会让人感受到一种温馨而自然的舒服之感。

三叶草：一种新的绿色色调出现，这便是饱和度极高的三叶草。这种高饱和度的颜色其实非常难以驾驭，但是并非不可驾驭。在家居环境中，这种颜色会让你瞬间感受到印第安人的热情和热带雨林的风貌。

深绿：深绿这种带来轻科技感的色彩已经成为了一种十分重要的商业时尚的色彩，可以预想的是在 2015 年的时装发布会上还会有更加抢眼的表现。

西柚：西柚作为一种新的色彩出现在米兰的秀场上，让它在 2015 年中的表现不容忽视。西柚色的特色之处就在于它极其富有表现力的红色，不似粉色腻人，也不似大红闪耀，多了一股知性和优雅，同时还拥有了一股少女的气息。

速品蓝：靓丽的水蓝色带着充满活力的运动或冲浪气息。速品蓝可以说最近才流行起来的，时尚而清新的色彩绝对是流行的不二之选，富有清新海洋之色的速品蓝也不负众望，为家居的陈设带来了一股清新的微风，感觉像是漫步在希腊的街头，张开双臂就可以拥抱海风。

靛蓝：靛蓝色已经实现普及，作为一个商业时尚的色彩，2015 年将继续走红。而很显然，这种商业又时尚的颜色会让家居时尚变得更为时尚。总之，靛蓝的颜色会让你的家居变得富有贵族般的奢华和摩登。

万寿菊：万寿菊一样的金色，与其他色调的黄金色有着不一样的内涵。它显得更为低调而内敛，同时也显得非常知性。在家居环境之中，这种颜色不会如一般黄色一样让人不能接受，因为它的色调较为低缓，让人享受，同时也不会太过喧宾夺主。

薰衣草：浪漫的薰衣草色仍将继续流行，淡淡的紫色让人感受到魅力。魅惑的紫色会让人心生压抑，但是这种淡淡的紫色会让人感觉到甜蜜和优雅。在家居中，如果你喜欢紫色，但是又不想显得太过气势逼人，那么就选用这种淡淡的薰衣草色吧，它肯定会让你的家变得耳目一新。

玛莎拉酒红（Marsala）：色号为 18-1438 的玛莎拉酒红，这一故事感成熟、韵味极强的色彩，象征着信心与持之以恒，最能反映出人们的渴望与感受，注定会为 2015 年加入浓郁的时尚气质。玛莎拉酒红，源自于意大利西西里岛经典葡萄酒的色彩。正如它的名字一样，玛莎拉酒红温暖而饱满，热情而优雅，美得令人心醉。这种颜色不仅十分吸引目光，还具有一股诱人的成熟韵味。在居室里大面积地使用它，可以营造出优雅的生活氛围，身处其间，能够体会拥抱温暖的感觉。

淡青（Lucite Green）：淡青纯净、清新、清爽，带给人思乡般的感觉。

冰川灰（Glacier Gray）：低调舒适的冰川灰代表着时间的永恒，安静得令人宽慰，平和得使人放松，虽然自身并不出众，但却具有极强的包容性，已经在多季的时装周中被捧为经典。

烤杏仁色（Toasted Almond）：温和的烤杏仁色给人温暖舒适、置身春夏天的感觉，就好像捧在手中的细沙一般，是人工味道最低、最贴近自然的选色。

勃艮第酒红：勃艮第酒红是红色的一种，因与法国勃艮第所出产的勃艮第酒颜色相似而得名，与栗色相似。

粉晶：代表舒缓、接纳和滋养。Coral Blush启发爱与接纳性，银色感代表改变的价值和接纳，而粉红则代表爱与温柔。两者结合成一种充满情感疗愈功能的颜色。

孔雀蓝：孔雀蓝是蓝色中最神秘的一种，几乎没有人能确定它正确的色值所在，是模糊色的一种，不同的人会对它有不同的诠释，代表的意义是隐匿 。在印刷领域里，这种颜色会和设想产生很大的误差。在精神领域里，这种颜色是遥不可攀的神界的颜色，是除了金银以外的一个特殊色。 它的色彩是模糊的，它象征着诡异的重生。它会以一种特殊的方式存在，在隐隐之中给人一种心理的暗示和神秘的力量。所以，它的意义是非同寻常的。

薄荷色：薄荷色是清凉主打色，占据着足够多的“地位”，流行趋势不可挡！

驼色：和红色、绿色等鲜艳的色彩一样，驼色同样来自于大自然，来自于苍茫的沙漠、坚韧的岩石……但有趣的是，这种来自于自然界的色彩却具有一种非常都市化的味道，驼色是淡定的，就像一杯恰到好处的清茶，酽而不燥，淡而有味，是搭配中令人放心的底色——平和宁静，但绝不乏味。

亮黄色：设计上没有花哨的装饰，亮丽的黄色一样能带来视觉上的冲击，温暖而亮丽的黄色墙面，或者是家中一件黄色座椅、黄色边桌、黄色灯饰，都能令仍有一丝寒意的早春空间变得缤纷绚丽。

红橙色：充沛橙色，将是2016年流行的主要颜色，看起来就像是在橙色里加了一些粉，能让设计的配色变得更为活跃和更具活力感。

太妃糖色：作为大行其道的复古色系中的一员，介于棕褐色与沙漠色之间的砖黄色（toffee、太妃糖色），在本季显得尤为突出。这一色彩有着20世纪60年代的波希米亚风格，同时也带有些许现代游猎风格的乐趣。

松柏绿：松柏绿，中国传统色彩名词，松柏叶的绿。深沉而旺盛的颜色，让整个色彩变得低调而飞扬。用其装饰室内，可以营造出复古的低调感。

鸽灰色：鸽灰色是一种柔软而有穿透力的色彩，低调而充满禅意。在北欧风格的设计当中，鸽灰色是非常常见的颜色，这种品质的色彩非常适用于时尚的设计当中。

月长石蓝（Moonstone Blue）：在冷色调里，月长石蓝是一种让人兴奋的新流行的时尚颜色。

麂皮色（Chamois）：麂皮色的灵感来自夏天山羊光泽的毛质。在春夏流行色中，麂皮色取代了低饱和度的黄色及浅卡其色，为新一季的中性风提供了更多选择。

复古靛蓝色（Vintage Indigo）：经历了褪色处理的复古靛蓝色将靛蓝色的饱和度降低，微妙的变化增添了些许的神秘色彩。

赭棕色（Sienna Brown）：红色与大地棕色的完美结合给传统颜色开辟了新的领域。

淡雾紫色（Mauve Mist）：淡雾紫建立在粉彩色彩的基础上，融合了更多的中性风格，它比灰色更暖，令人愉悦却也不失格调。

空间调色盘

Platte

项目名称：心房
设计公司：布拉尼设计
设计师：布拉尼、德西斯拉娃

Project Name: Life in Expressionism
Design Company: Brani & Desi Studio
Designer: Branimira Ivanova, Desislava Ivanova

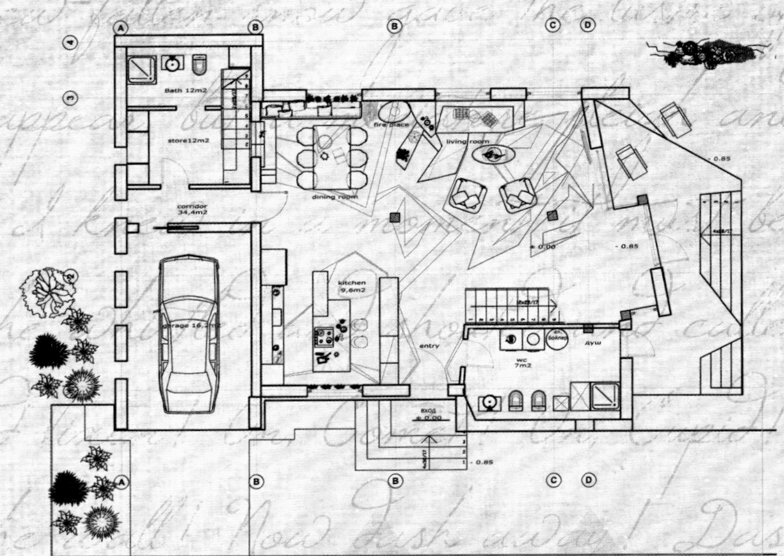

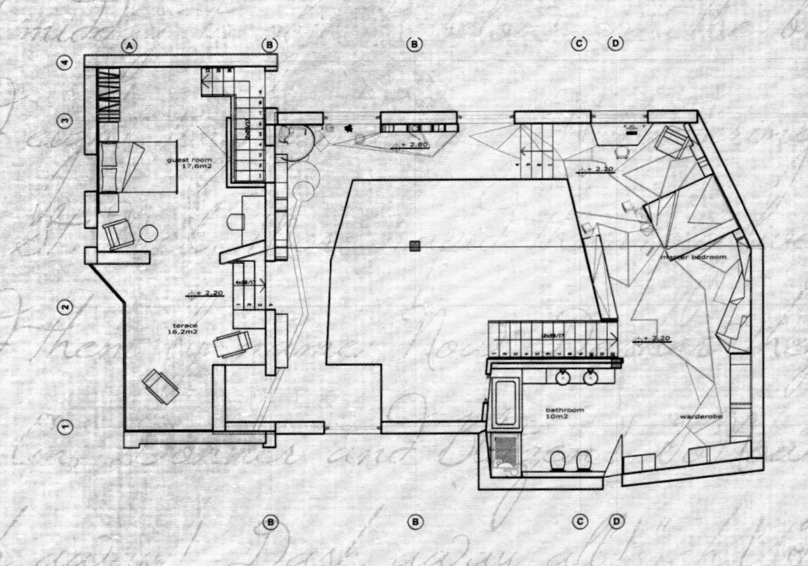

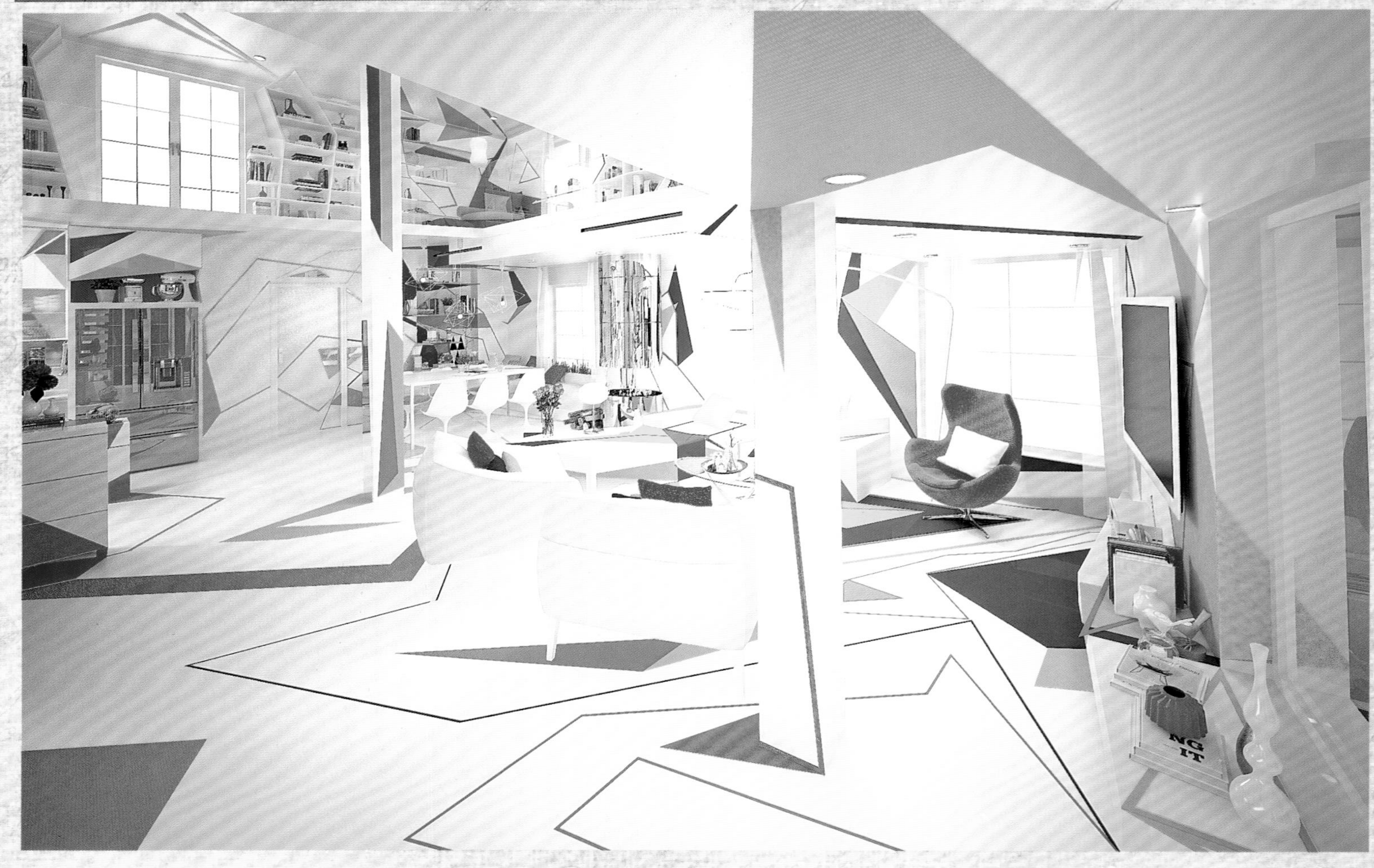

造型与色调的运用，塑造出温暖、舒适、自信、快乐的创意空间。白色的纯洁闪现于其中，不同的造型与色度似乎出现了一丝停顿。

厨房明亮而富有能量。红色、橘黄色带来的是创意、力量的动感。白色的清纯与温柔平衡着两种强烈的色调。餐厅里的蓝绿以其宁静抚慰着人的灵魂。

客厅里虽然着以绿色，但却稍有差别。红色的沙发、抱枕以及楼梯旁的造型以同样的绿色共同打造出一个适宜放松、交谈的天地。

穿过客厅旁的楼梯，便到达了卧室。卧室分为几个区域，有的便于休息，有的便于工作，而有的则便于阅读。蓝色、紫色及黄色给人以放松的清新感。

书房及健身房位居二楼。这里是一个用情感体验而不是用身体进行体验的空间。

The project employs shapes and colors to evoke emotions ranging from feelings of warmth and comfort to feelings of confidence, enjoyment and creativity. The white color is the active pause amongst the shapes and tones.

The Kitchen is bright and energizing for the users. The red and orange colors feed them with the energy of

Elliott Erwitt Snaps
SALAD + DINNER
JUICY DRINKS

NEW YORK

creativity and power. The purity and softness of the white balance the two strong colors.
The blue and green colors in the dining room impact on the soul with its energy of calmness and relax. The acute forms of the colored shapes balance the cold colors by giving warmth to the room.
The living room is the main room for relax and conversations. The green is the most pleasant for human eye. We use few nuances of the green. The red color of the sofa and the pillows, and also of some shapes in the stairs zone balance the green as its complementary color.
The bedroom is reachable through the stairs in the living

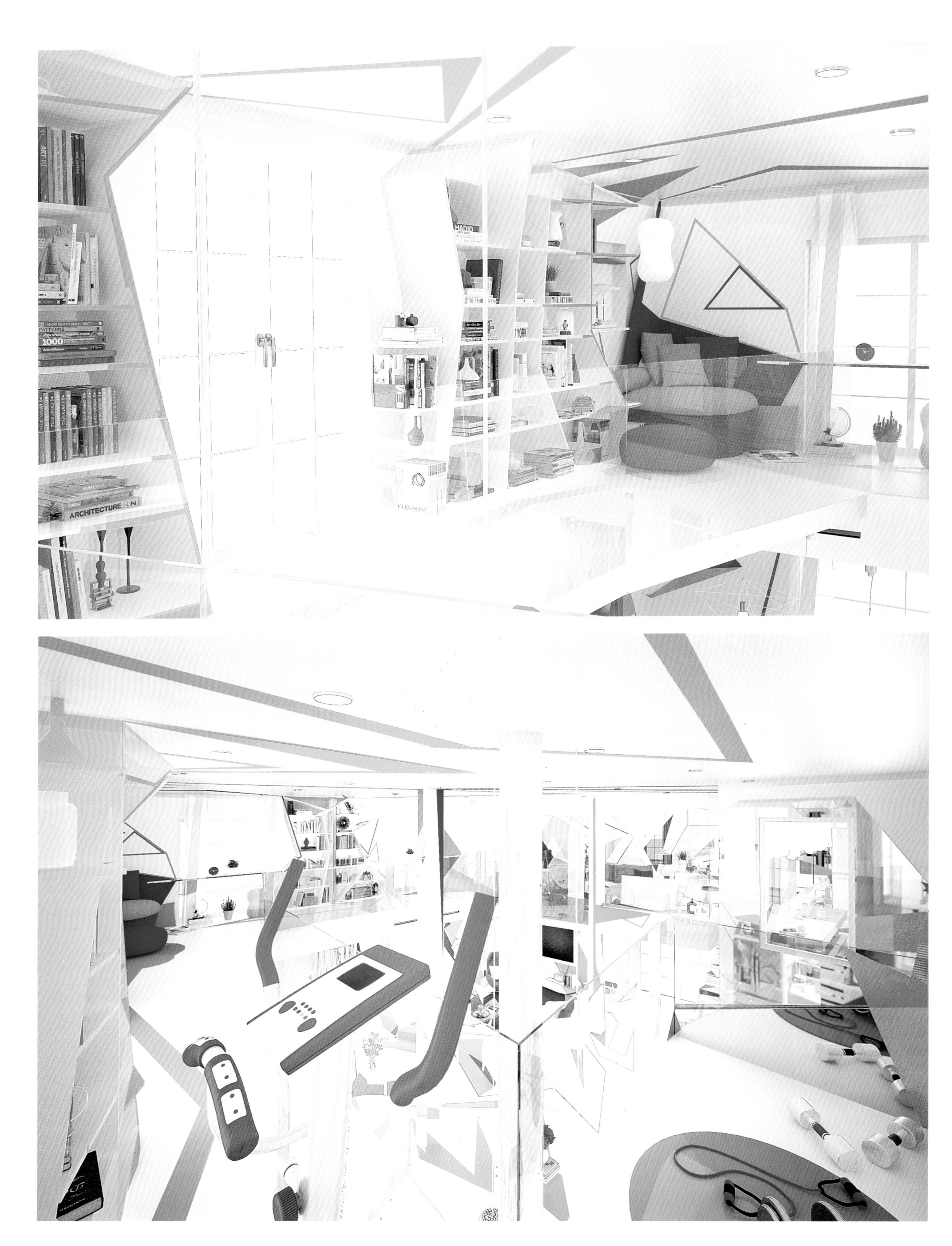
ARCHITECTS TODAY
PATTERNS
1000
ARCHITECTURE
HADID
GERVASONI

room zone. The space is divided into a few zones, where one can sleep, work, read etc. The blue, violet and yellow colors and the forms, which they fill, are creating a relaxing and fresh atmosphere. The library and the fitness zone are on the second level. To live in this place is one emotional experience rather than physical reality.

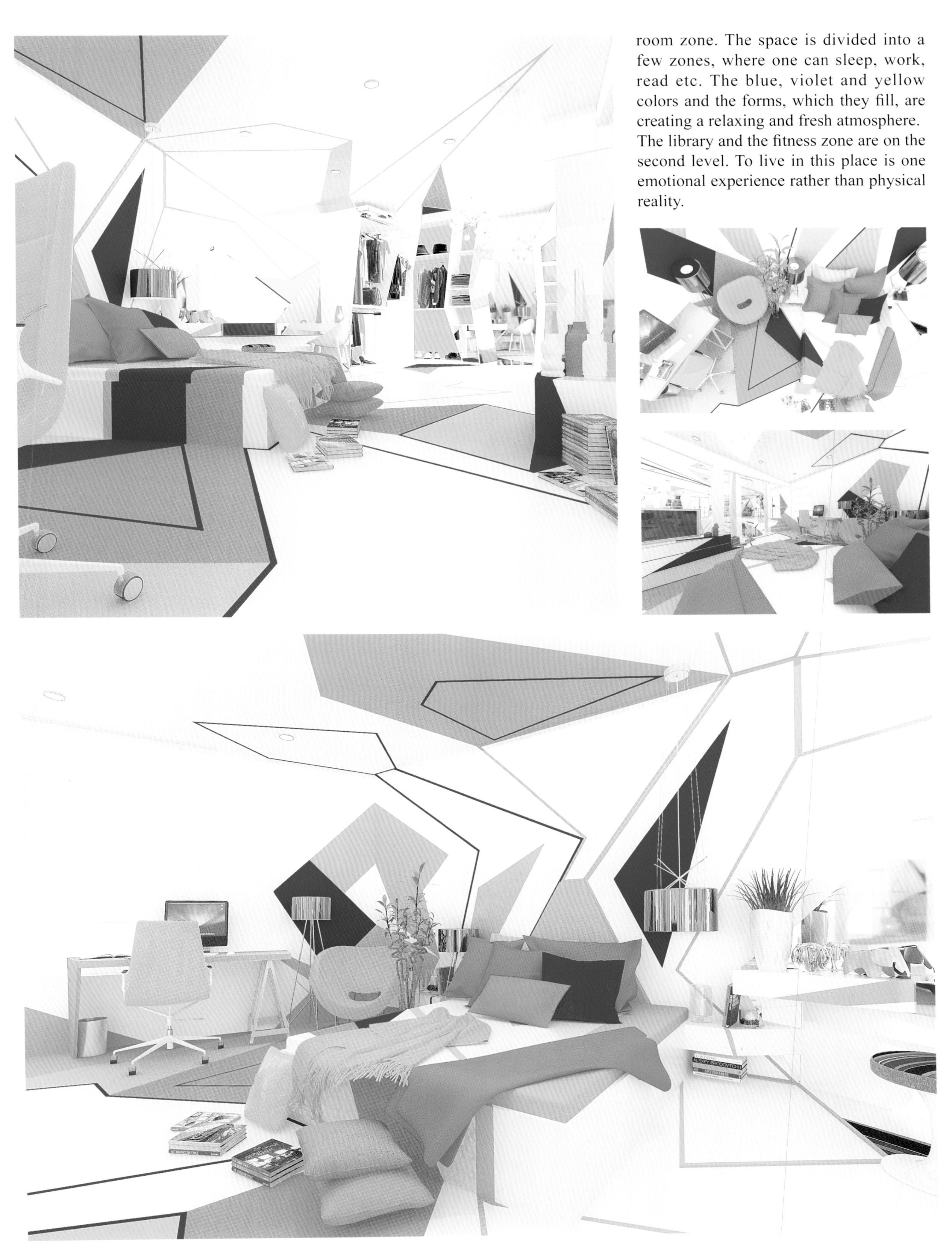

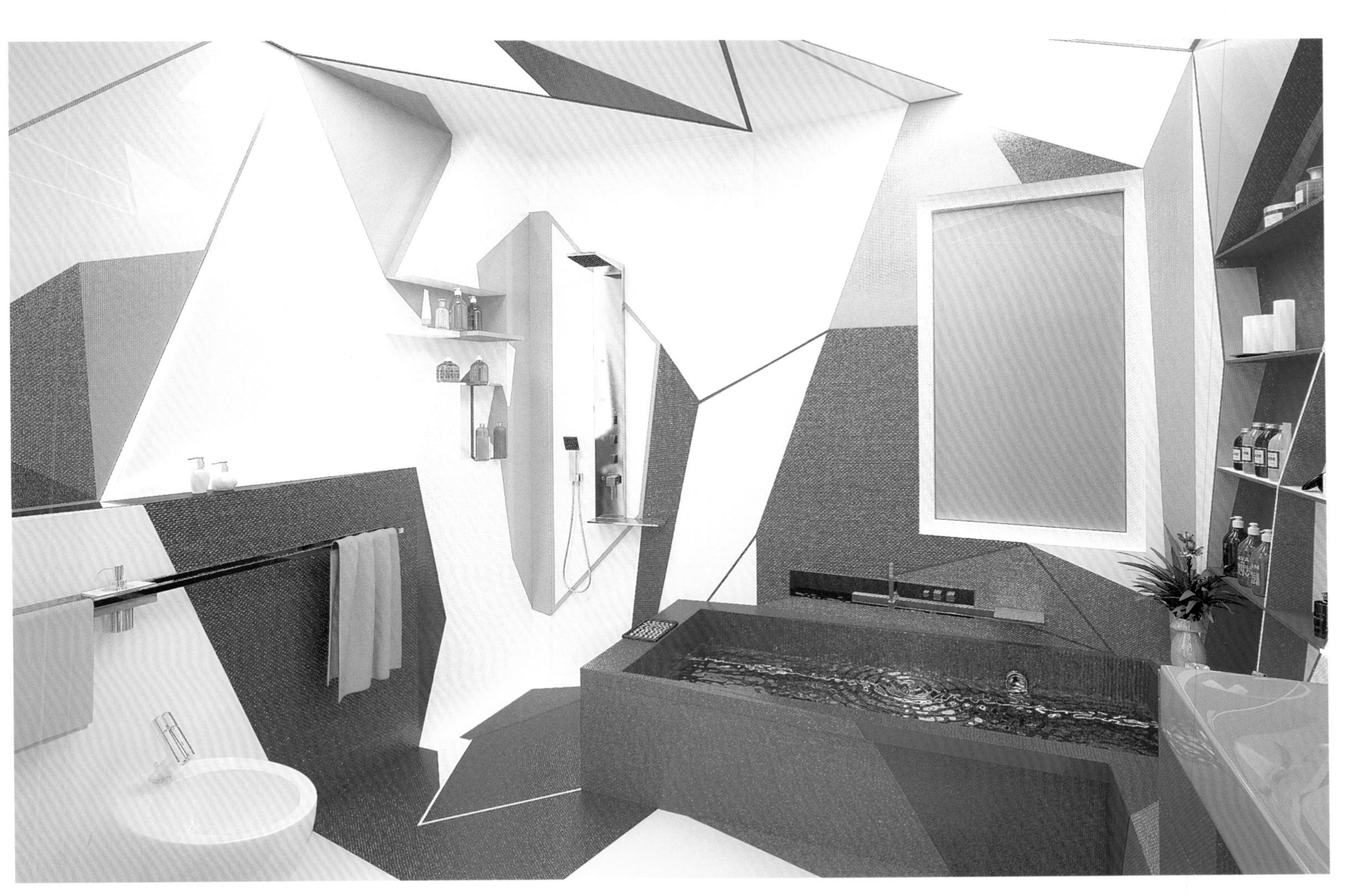

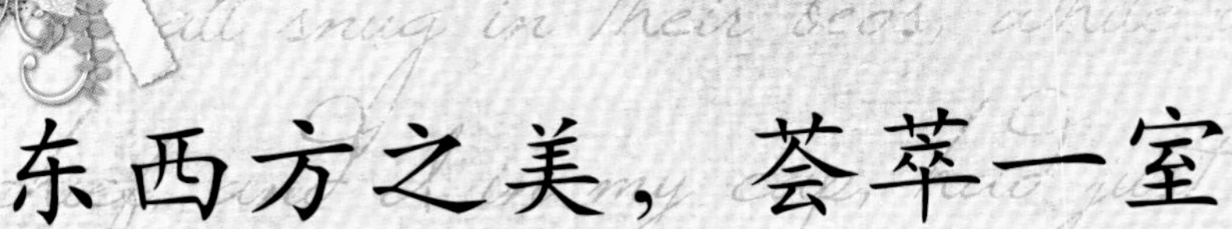

东西方之美，荟萃一室

Into One, Aesthetics of the East and the West

项目名称：新加坡顶层公寓
设计公司：发明设计
设计师：尼基基
摄影师：乔安

Project Name: Singapore Penthouse
Design Company: Design Intervention
Designer: Nikki Hunt
Photographer: Jo Ann Gamelo-Bernabe

黄金地段、美好的城市景观、朗阔的空间视野是影响业主购置此处房产的主要因素，但1 067平方米的空间几乎毫无个性可言。
经过设计，原来的电梯井给人一种视觉上的拉伸感。不仅精致、迷人，更让其天生具有一种雍容华贵的感觉。主入口所用壁板以黄色玛瑙石作为装饰，一种温馨的欢迎质感油然而生。
曲面的岛柜如同吸引人的雕塑，吸塑亚克力制作，外面以同样的黄色玛瑙石作为装饰，界定着开放式厨房与客厅之间的界限。岛柜上，悬挂着三件套的木皮吊灯，把自然元素从室外引至室内。
卧室洋溢着现代的浓浓中国风。定制的丝绸墙纸、中性色调的全套餐饮家具、边柜、窗帘、地毯等虽然相互映衬，但闪着最耀眼光芒的依然还是墙上那纤细、娇柔的丝绸。八角形的餐桌、展示柜，高光白色亚克力饰面，令人想起东方的漆器。装饰感很强的天花设计延续着客厅墙上别致的八角窗，增强了空间的凝聚力。
主卧显得特别沉静，典型的日式风格。珍珠白的墙面，高质感的草绿色面料，偶有木质模型间入。日式梁柱的做派。纸糊木框的滑动门、宣纸壁面的滑动门引向化妆间及浴室。抬高的床立于空间，看似"另类"，实际则很和谐。床体如同宣言一般，松绿色的软装大胆、清新，给人一种意想不到的融合感。
各式各样的意象以两种风格作为背景。原本毫不起眼的空间就这样给人留下前卫、优雅的印象，兼具东西方之美。

Location and view were key factors that influenced the homeowners' decision to buy this new 1,067 square meters penthouse located on Orchard Road, but despite its prime location and stunning city views, the penthouse lacked soul, character and personality.

The original private lift lobby was re-designed to heighten the sense of arrival and convey a luxurious setting infused with a subtle touch of glamour. The panels around the main entrance were replaced with honey onyx, an exquisite stone whose incredible depth of color radiates a warm welcome.

Separating the open-concept kitchen from the adjacent living room is a curved island bench that looks like a piece of modern sculpture made of thermal-formed solid surface acrylic, clad in the same onyx as the lift lobby. Suspended above the island is a trio of wood veneer pendants that incorporate a natural element into the room.

The dining room is a contemporary re-interpretation of Chinoiserie style. The biggest design statement has got to be the bespoke silk Chinoiserie wallpaper. This is complemented by a neutral ensemble of dining furniture, sideboard, curtains and rug to give the wallpaper the attention it deserves. The octagonal shapes of the bespoke dining table and display cabinet, both in white, high-gloss acrylic finish reminiscent of oriental lacquer ware, and the decorative ceiling design draw reference from the octagonal window in the living room feature wall to create an overall sense of cohesion throughout the interior.

The master bedroom evokes a serenity that is distinctly Japanese. The walls are wrapped with a white pearlised, highly-textured grass cloth, inset with a timber molding design that again draws reference from traditional Japanese post and beam architecture. Shoji screen sliding doors with rice paper panels lead to the dressing rooms and bathrooms. The only element that seems a tad out of place is the elevated bed but this is for good reason. The bed as a statement piece whose turquoise upholstery is bold and fresh and creates an unexpected juxtaposition.

By drawing upon various symbolic elements in two contrasting styles, this nondescript Penthouse has been transformed into a strikingly bold, yet stylishly elegant home that reflects a masterful blend of Japanese and Chinese styles with a contemporary twist.

JAPAN

源于沙漠的灵感，融入静美居庭

Desert Inspiration in Courtyard

项目名称：亚历桑拿酒吧
设计公司：贝克汉姆室内设计
设计师：朱莉娅
摄影师：埃里克

Project Name: Nolan Arizona Bar
Design Company: Buckingham Interiors + Design LLC
Designer: Julia Buckingham
Photographer: Eric Hausman

设计师朱莉娅的设计灵感源于自己曾经看到过的很多关于沙漠的照片。照片的色调、质感对设计师内心的触动，以及业主渴望奢华的环境中有一种宁静、温柔的气氛，最终促成了本案的设计。

客厅里孔雀蓝衬托着柔软的面料。墙面所用的定制预制木质建材、嵌入式架子、壁炉的出气孔是色系的调节器。中世纪的灯盏、名师亲手打制的椅子让空间多了别样的性格。

厨房除了复古式的吧凳，还有木材、大理石拼花打造的岛柜、新式的慕拉诺灯具。窗户除了装饰着定制的几何推拉板，还运用着闪闪发光的金叶子。

浴室里，从天花到浴盆以定制的渐变波纹瓦作为特色。海草纹饰的地板以“人”字形图案作为铺排。金属的“V”字形支架支撑着定制的梳妆镜。

主卧尤为引人注目的是临窗的土耳其式沙发床。古色古香的法式太师椅装饰的却是现代的风格。水晶吊灯则以多面形状出现。不同寻常的定制边柜出自旧金山艺术家之手，上面绘有金门大桥的剪影。

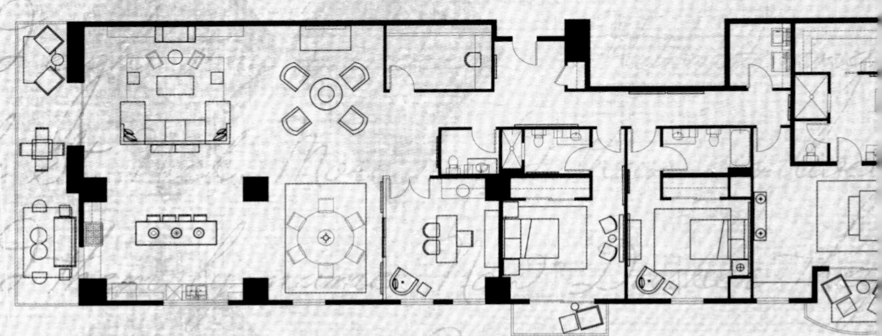

办公室以古典式、中世纪的家具作为铺陈。盥洗室以皮埃尔·福雷（Pierre Frey）的立体墙饰作为美化。白色的垂直面体上绘有黑色的玫瑰。黑白二色的大理石梳妆镜、照明烛台其实是意大利的著名家居品牌 Fontana Arte。

For this project Julia was inspired by the colors and textures of the desert seen from many of the picture windows of the luxury condominium as well as by the client who requested a soothing and feminine environment.

The living room is a lively combination of peacock colors and soft fabrics. To give the room additional structure Julia added custom millwork for walls, built-in shelves and installed a vent less gas fireplace that changes color. Mid century lamps, Pedro Friedeberg hand chairs and an antique Mid West barn crest lend additional character.

The kitchen offers a host of textures including retro counter stools that combine wood and rush, a gold, copper and marble tile mosaic for the dining island, vintage and new Murano glass light fixtures and custom geometric sliding panels used as window coverings made from gold leaf.

The bath features custom rough-tumbled ombre tiles that flow from the ceiling to the bath juxtaposed with seagrass floor tile laid in a herringbone pattern. The custom vanity sits on metal v-shaped legs.

The master bedroom boasts custom-designed elements from the Turkish daybed set in the window. Additional highlights include a pair of vintage French fauteuils upholstered in modern ombré fabric, a faceted crystal chandelier and an unusual side table custom-made by a San Francisco artist using pieces of the Golden Gate Bridge.

The owner's office combines neoclassical and mid-century furniture and art including a 19th century bureau plat. Finally, the powder room is embellished by a Pierre Frey three-dimensional wall covering with a black rose motif on white, a black and white marble vanity and a pair of illuminated sconces in the manner of Fontana Arte.

畅享 70 年代的艺术氛围

Art Ambience from the 70s

项目名称：海晏阁
项目位置：法国安提布
建筑师：卢克
设计师：基里尔
摄影师：西蒙

Project Name: Seafront Villa
Location: Antibes, France
Architect: Luke Svechin
Designer: Kirill Istomin
Photographer: Simon Upton

“海晏阁”建于 20 世纪 80 年代，由俄罗斯移民建筑师安德烈所建。业主购买时，房子状况非常不好。几经周转，他们找到了安德烈的儿子同为建筑师的小安德烈，希望在他的帮助下，对空间重新进行修整。在小安德烈的主持下，楼层面积得到了扩大，卧室增多，而同时空间比例没有改变，景窗、气派的楼梯依然得到了保持。整个风格如同一个现代主义的加州别墅。设计师基里尔以 20 世纪 50—70 年代的设计精髓为灵感，对空间进行了室内设计。除了威廉·韩思（William Haines）的作品，大量的家具铺陈虽出自基里尔之手，但无不是 20 世纪 70 年代的风格，如软装的家具腿部支撑、漂白的核桃木器、花饰等等。还有些器物，古色古香，别有韵味。空间无任何过度装饰。放眼望去，色彩在一点点消退，与外面的盎然生机形成鲜明的对比，但给人的感觉却是安详、宜人、放松。

This elegant multi-level house on Coted' Azur was built in 1980s by Russian immigrant architect Andre Svechin. When KirillIstomin's clients bought the house the condition was already not so good. They chose the most intelligent way to reconstruct it: invited Andre Svechin's son, who is an architect himself, to do that. Svechin junior increased floor space of the house adding more bedrooms but keeping the proportions and house's main attire – panoramic window sand spectacular staircase. The residence resembled a modernist Californian villa. So Istomin, who was invited to decorate it, chose 1970s and William Haines' interiors as inspiration: the spirit of California of 1950s and 1970s, decorative but rather modest style, several direct quotes from William Haines' actual projects. Along with armchairs and so far by William Haines Designs, there are numerous object of Kirill Istomin's design, in the 1970s style (upholstered legs, bleached nut wood, flower prints), and vintage accessories. Istomin's idea was to use faded colors, without over-decorating the house. The interior is contrasting the vivid nature outside, creating a truly relaxing and soothing atmosphere.

浓墨重彩

Thick and Heavy in Hues

项目名称：色
设计师：尤里

Project Name: Color Object
Designer: Yuri Zimenko

“色”空间共有面积130平方米，设计师尤里主持设计。业主为年轻人，内部现代、色调鲜艳。虽说是色泽繁多，如深的蓝、翠的绿、柠檬的黄等，但高科技的手段却把不同色系平衡地融于同一空间。

公共区域有大厅、客厅、餐厅、客用卫生间。隐私区则有带有休息间的主卧。主卧配有飘窗、卫生间。著名波普艺术家谢尔盖的肖像、亚历山大的绘画作品悬挂于其中，正是年轻业主的最爱。

This modern colorful apartment (130 square meters) in Kyiv is designed by Yuri Zimenko for a young man. The main concept is based on high-tech interpreted in multicolor way, that's why the main task was to keep the complicate balance between such different shades: deep blue, emerald, lemon, coral...

The public zone includes the hall, a drawing room, dining room and a guest-bathroom, while private part consist of master-bedroom with the sitting area in a bay window space and master-bathroom. The special atmosphere is created by the art-works like the images of the owner's favorite actors created by Sergey Grinevichin a pop-art style, or Alexander Nekrashevich's painting with a poodle.

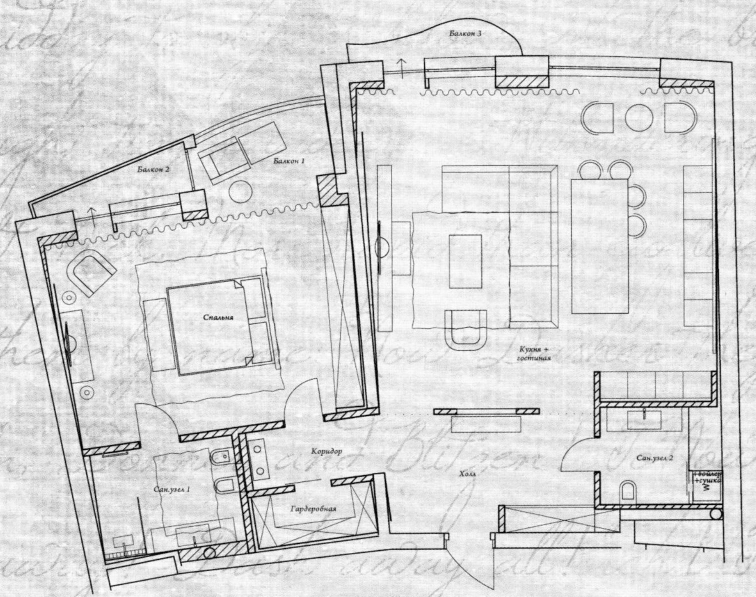

ITALY

人生旅途，收藏快乐之家

Life Journey to Accommodate Dwelling Happiness

项目名称：快乐之家
设计公司：维克设计
摄影师：维姬

Project Name: The House of Joy
Design Company: Vick Vanlian Design
Photographer: Vicky Moukbel

业主希望能有一个度假之家。打开门，给人的感觉便是“清新、快乐、开阔”。多彩的动感如同业主给人的感觉。“快乐之家”的名称原本寓意于此。

整体上如同风格、质感、色调之间的相互对话。自玄关起，客厅便给人一种波浪般的感觉，动感与旋律尽在其中。家庭电视室、主客厅空间的东方式遮窗格栅界定空间的同时，更有一种鲜明的对比。

彼此之间的界定精心考究。随着步履的穿行，各色系清晰可见。各经典的家具铺陈混杂其中，彰显本案设计的风格。明亮的色调，自然的图案、纹理，尽在本案。即便各处墙角也给人一种阳光的能量。

波普艺术、花艺方毯、几何图饰增加细节个性。伴着优雅，各元素和谐于彼此之间，传达着一种自由、快乐之感，也表达了业主对生活的向往。

The owners wanted a vacation house, and the first thing you notice in this apartment, as soon as you open the door, is its freshness and the fun, airy feel of the space. It is colorful and vibrant and essentially resembles its inhabitants. And therefore, it was given the name: House of Joy.

The concept here was to play on the dialogue of styles, textures and colors. Starting from entrance, the reception area wall is covered with wave like textures that create movements and add dynamism to the space. This is then contrasted with a subtle Oriental Moucharabieh that separates the TV family room, and the main living space.

Space transitions are studied carefully and made visible by the use of colors as we go from one space to the next. Mixing classically inspired furniture, made the Vick Vanlian way; and Contrasted with bright colors, nature inspired patterns and textures; every corner of the space sends positive vibes and energy.

Pop art works, rugs of floral or geometric motifs, personalized added details, all masterfully orchestrated together with elegance to convey the free, life loving, joyful personalities and lives of the owner.

FASHION
VERSAILLES
سكويرت
ولويز أن
أفتر-بورست!

THE SOCIAL BOND

MY SHOES
MY SHOES

FASHION
Dior
25,000 YEARS OF Jewelry
JAMES BOND 50 YEARS OF MOVIE POSTERS
THE SOCIAL BOND
La Photographie
Paris
VERSAILLES
FASHION ADDICT
Pucci

折中风格，不改快乐本色

Eclectic yet to Be Gay

项目名称：好莱坞顶层公寓
项目位置：美国加利福尼亚
设计公司：艺术线工作室
设计师：马克西姆
摄影师：蒂格兰

Project Name: The Penthouse
Location: California, USA
Design Company: Studio Artline Inc
Designer: Maxime Jacquet
Photographer: Tigran Tovmasyan

“好莱坞顶层公寓”如仙境一般，身置于其中，自然是无尽的想象。虽然是折中的风格，但时尚的背后依然有安详，现代的幽默背景中有壁画。经典的卡通给人一种动感，并伴有一种闪烁的霓虹。古老的经典卡通映衬着奢华。除了超大的金宝汤业的罐子，还有古董的自行车及唱片电唱机。这里有时尚、快乐、复古的铺陈，这里也有设计师真正的天分展现。

The Penthouse, a wonderland of trends and periods captivates its guests and transports them to a medium beyond ones imagination. His eclectic design is extremely fashion forward and the atmosphere is serene and decadent. He incorporates modern humor with murals spoofs on classic cartoons, ignites an energetic surge with colorful neon, and splurges with designer throw blankets by Hermes and Louie Vuitton. Its luxury at its best with a twisted vintage flair, with accents like an oversized can of Campbell's soup, and an antique bicycle and record player. This bazaar of fashion, fun and retro time pieces, is masterful work of a true designer genius.

绿茵环抱，彩色之家

A Colorful Home Among the Greenery

项目名称：热带别墅
设计公司：发明设计
设计师：尼基基
摄影师：乔安

Project Name: The Tropical House
Design Company: Design Intervention
Designer: Nikki Hunt
Photographer: Jo Ann Gamelo-Bernabe

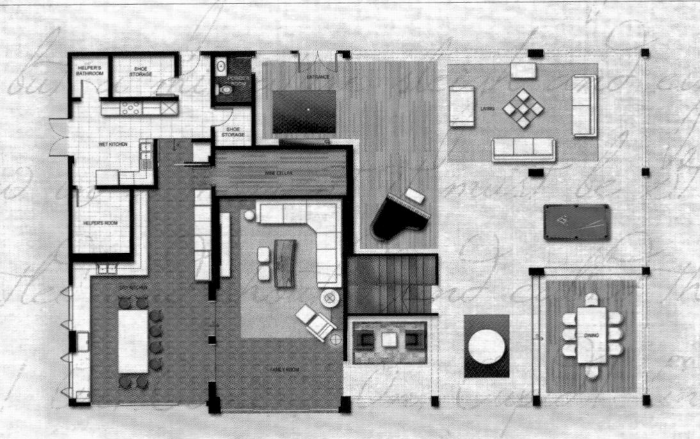

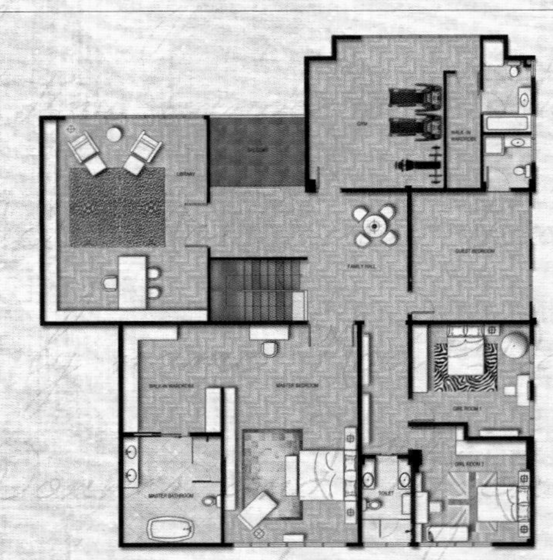

新加坡"热带别墅"享尽自然之青睐，拥有开放、开阔的地基，但设计丝毫不改家居空间之本质，内部无不彰显生活之意义、人生之快乐。虽然位于城市核心地带，附近为新加坡的一块公有地块，任何角度观望都是茂盛、郁郁葱葱之景象。设计与本案的意义，便是充分利用得天独厚的地理条件。

各空间平静，无物品拥挤之虞。但是鉴于空间之广阔，没有足够的家具铺陈，空间难免显得有些清冷。为了避免这种情况的发生，具有纹理的木材、粗糙切割的石块，有褶皱纹理的皮革、牛皮运用其中，柔化着其中的硬性条纹，填补了空间视觉上的空白。内部空间的线条如同家具本身给人一种坚实的感觉，强化着所在基地赋予人的力量及固有的永恒。

一楼的空间，很大一部分用于厨房、家庭活动室。整体给人一种开扬、明亮、通透的感觉。正好让业主发挥大厨的特长。休闲、放松的环境里满足了业主的个人爱好。各个房间的艺品无不经过精心挑选，大胆、前卫、精彩。幽默的手法，杜绝了一切"做作"的意象。好一个"热带别墅"，反映个人品位与喜好的同时，依然给人一种优雅、简约、温馨、热情、快乐的感觉。

The Tropical House in Singapore is an expensive open plan property which, first and foremost, was to be a family home, full of life and fun. Although located in the heart of the city, it is bordered by state land and has lush, verdant views from every aspect. Design Intervention's task was to capitalize on this vista at every opportunity.

The client wanted a serene but happy home and so great care was taken not to crowd the rooms with too many items to ensure a calm, clutter-free environment. However, as the rooms are so large there was a risk that they could appear rather cold. Avoiding this, textures were added such as wood grain, rough cut stone, distressed leather and cowhide to soften hard lines and visually fill the spaces. The lines of the interior architecture as well as the furnishings are strong, almost chunky, to invoke a feeling of strength and permanence and match the scale of the property.

A significant amount of floor space was allocated to provide a kitchen and family area that was spacious, bright and airy. This perfectly suited the client who is a keen cook and enjoys entertaining but in an informal, relaxed style. Artwork choices throughout each room are bold, bright and colorful and inject a touch of wit in this unpretentious family home. Reflecting the personality of its owner, it is elegant, yet uncomplicated, warm, welcoming and lighthearted.

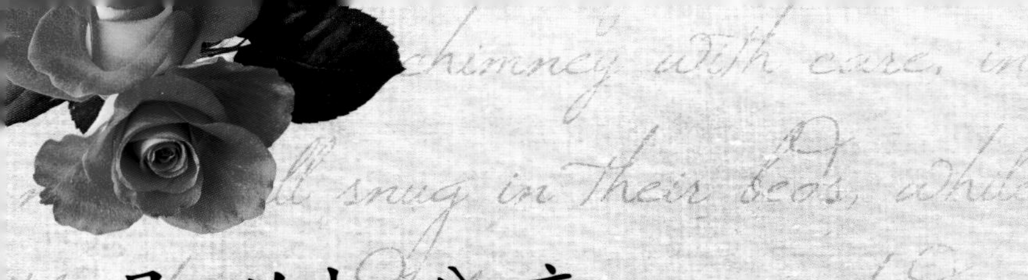

即兴切分音

Improvised Syncopation

设计公司：城市室内装修设计有限公司
设计师：陈连武

Design Company: Chains Interiors
Designer: Chen Lianwu

本案开放式的空间设计，让工作、用餐、烹饪、视听娱乐及聚会聊天等活动可以随时随地同时重叠进行，满足家居生活与工作结合的新生活趋势!
室内黑木地板延伸至户外，除了让视觉得以从室内延展至户外之外，更是休闲与工作的结合。

即兴切分音

因业主热爱古典乐，设计师在其生活的空间中置入音乐性，将音符及节奏转化为空间的线条与色彩。
以《幻想即兴曲》的创作手法为发想，在开放空间中，运用大面积的白色背景墙面，以不等距的金属条切分线，不断重复包围，有如偶发的切分音伴奏和弦，带出随性轻松的空间调性。
活动家具及主墙面，以个性强烈的色块与图案，带出幻想浪漫的主旋律。同时也与条状镜面背景材质互相交织成一首《幻想即兴曲》。
以肖邦《幻想即兴曲》的曲式架构为灵感，发展出对应于空间中的音乐性。

导奏

空间的入口（玄关、隔屏），大面紫色墙与平行的垂直线条，带出空间的迎导前奏!

A段主题

公共空间中（餐厅、客厅），不等间隔距离的垂直线条，犹如肖邦的《幻想即兴曲》，以左手连续琶音贯穿全曲，赋予空间速度与张力；而空间中央的红紫色调，犹如右手不同风格的主旋律，在四周的垂直背景线条中被衬托出来。

B段主题

主卧空间横向水平线条，犹如《幻想即兴曲》中的抒情旋律，赋予空间柔和缓慢的步调。

尾奏

儿童房空间以横向为主，垂直向为辅的线条，犹如A段主题与B段主题的穿插互换，将儿童房的活泼性格凸显出来。

To satisfy working and living at home, open space design would be the best solution for this new living trend. To make working, dinning, cooking, entertainment, and social chatting all together and happen anytime.

Black wooden floor extends from inside to outdoor. Besides the visual extension, it also combines leisure and work.

Improvised Syncopation

A sense of musicality is installed into the living space for the apartment owner with a love for classical music. It transforms notes and rhythm into spatial lines and colors.

Taking the basic idea of improvisation in music as a starting point, the open spaces in the background make wide use of white wall areas. Metallic stripes with irregular spaces between them syncopate the spaces and at the same time encompass them in a repeated manner. This has an effect similar to spontaneous syncopations in musical accompaniment or to the strings of musical instruments. They thus create a spatial atmosphere of calmness and lightness.

Movable furniture and the main wall with their characteristic colored areas and patterns convey the main melody of fancy romanticism.

Inspired by the composition of Chopin's *Fantasie-Impromptu*, musicality evolves that corresponds with space.

Prelude

Entrance to the rooms (partition in the entrance area): an extensive purple wall and vertical and horizontal lines mark the beginning of a prelude designed to welcome the guests!

Leitmotif A

In the public space (dining room, living room) vertical lines with irregular spaces inbetween produce an effect similar to Chopin's *Fantasie-Impromptu*: the continuous arpeggios of the left hand run all the way through the whole piece and lend it spatial velocity and tension. By contrast, the red and purple tones of the central space have the effect of the leitmotif of the right hand. It is characterized by different kinds of styles and stands out from the surrounding vertical lines in the background.

Leitmotif B

Horizontal lines run through the space of the main bedroom, evoking the lyrical melody of the improvisation. They give the room a soft and quiet character.

Postlude

Horizontal lines dominate the children's room, complemented by vertical lines. They reflect the intertwined character of leitmotifs A and B and thus give expression to the liveliness of the children's room.

散发柠檬气息的高阔明朗之居

Bright Residence Above Lemon Air

项目名称：亚细亚
设计公司：S&K 室内设计

Project Name: Asian
Design Company: S&K Interiors

只要去蒂娜、特里的家里，客人对于玄关的印象总很深刻。开放、通透，如同主人给人的感觉，是此处空间给人的第一印象。两层的空间开有高高的窗户。高高的窗上安着的楣窗，使空间更加“高大上”。开往室外泳池的门户扩大了娱乐室的面积，便于大型派对的举行。黄色的墙面，间以白色，给人一种阳光轻松的感觉，客人即刻便有一种宾至如归的感觉。
业主夫妇选择 S&K 为其设计该空间的玄关。于是，墙上中性的象牙白、黄灿灿的金色立刻让此处鲜活起来。蓝绿色的沙发配上奶油色的后背。三三两两的靠垫是男主人亲自挑选，搭配着空间的蓝、绿、红、黄。其中几个红色的靠垫印上了波罗的设计，传统的热情好客尽在其中。镶嵌式的书柜背景着以红色，凸显着柜内的摆设与艺品。
天花上两座水晶吊灯，高高的悬挂，低调、内敛的姿态引人注意。一旦降低高度，却发现原来是空间优雅的代言。吊灯上的每盏灯刻意加上灯罩，与其他房间红色的色调相互映衬。

Guests who visit Tina and Terry Crouppen are greeted by their cheerful welcomes and the open, airy entrance seems to echo their sentiments. The two-storey room, with tall windows topped by fanlights along one wall, adds to the feeling of spaciousness. Doors leading outside to the pool area provide options to expand the entertaining area for large parties. The yellow walls trimmed in white add a sunny atmosphere to the space, where people instantly feel at home.
The Crouppens chose S&K Interiors to help them achieve a new look for their entranceway. S&K began by painting the walls, which had been a neutral ivory, in a golden yellow, immediately brightening the room. They created two comfortable seating areas, using a blue-green sofa with cream leather backing that Tina had recently bought and adding accent pillows in matching blue greens, reds and yellows. Several red pillows feature a pineapple design, a traditional symbol of hospitality.
An additional red accent was added by painting the back wall of a built-in bookcase. The wall provides a dramatic backdrop for the art objects displayed in front of it.
Two chandeliers already hung from the ceiling, but up so high they were mostly unnoticed. Shirley lowered them, making a statement of elegance. She added shades for each of the lights on the chandelier, coordinating them with the rest of the room by their red color.

WONDERS OF THE WORLD
WONDERS

W 公寓，极尽简约，品质至上

W Apartment: Utmost Simplicity, Quality First

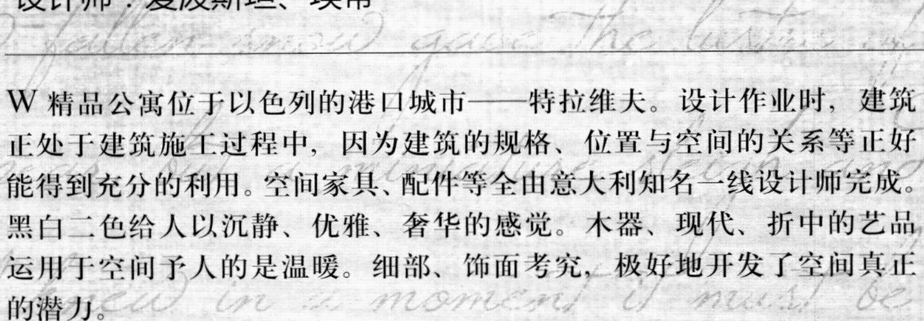

项目名称：W 精品公寓
建筑师：阿齐扎
设计公司：安多工作室
设计师：爱泼斯坦、埃蒂

Project Name: W Boutique Apartment
Architect: Aziza
Design Company: Ando Studio
Designer: Epstein Anna, Azougy Ety

W 精品公寓位于以色列的港口城市——特拉维夫。设计作业时，建筑正处于建筑施工过程中，因为建筑的规格、位置与空间的关系等正好能得到充分的利用。空间家具、配件等全由意大利知名一线设计师完成。黑白二色给人以沉静、优雅、奢华的感觉。木器、现代、折中的艺品运用于空间予人的是温暖。细部、饰面考究，极好地开发了空间真正的潜力。

The apartment is located in Tel Aviv at a luxurious residential Project called "W Boutique". This apartment was designed for a private client. The design purpose was to show the full potential of the apartment (size, location, relations between spaces etc.) since the apartment is still in construction stage. The design is very modern using the first line Italian designers' furniture and accessories. The color scheme is quite monochromatic using black and white as basic colors to inspire calmness, elegant and luxurious atmosphere. To "warm" the feeling in the apartment we used wood combined with contemporary/eclectic art pieces. A great effort was done on every detail and attention to every finish material in order to reflect the true and best potential of all the spaces.

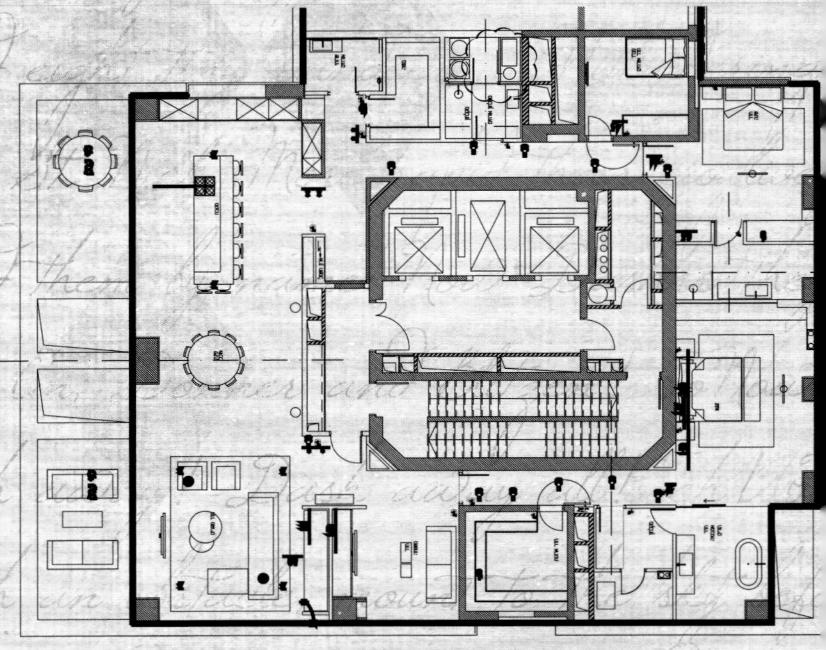

TOM
FORD

橱柜概念隐藏功能分区，小空间也有大享受

Cabinet to Hide Segments; Space Small, Enjoyment Great

项目名称：书苑
项目位置：乌克兰基辅
设计师：马赫诺
摄影师：安德鲁
面积：50 m²

Project Name: Open Book
Location: Kyiv, Ukraine
Designer: Sergiy Makhno
Photographer: Andrey Avdeenko
Area: 50 m²

空间虽说小巧，但业主还是希望能融入所有功能，如客厅、厨房，甚至步入式衣柜、书房等。为了满足业主的要求，本案50平方米的空间运用了"橱柜"的设计理念。

内部如同一组橱柜隐于滑动的玻璃推拉门之后。各空间的安排如同火车的卧铺。客厅位于玄关。衣柜、浴室紧随其后。浴室里马桶、花洒、浴缸一应俱全。书房紧邻浴室，墙面上布满了书架。紧靠浴室的是步入式衣柜。而最后的卧室则如同带有卧床的小盒子。一旦门楣洞开，生活空间则倍显深度。

开放的客厅区设有厨房。厨房里配备着一个修长的岛柜。岛柜周围正好是餐饮的区域。其中一个宽大的沙发正好位于家庭影院的对面。沙发是罗奇堡（Roche Bobois）的品牌，恰当的位置界定着其他内部的休息空间。墙面富丽的紫色辉映着沙发的橘黄色，同时呼应着餐饮区家具大师康斯坦丁·格里克（Konstantin Grcic）设计的椅子。

通过纹理表面的运用，色度并非仅仅给人以惊喜。白色的定制柜组亚光表面。厨房、浴室的地板有黑色的大理石，还有自然的实木。虽然迥然不同，但却相得益彰。客厅里、卧室里平面的板面以本案设计制作的3D立体面材作为替代。多形式的照明伴着嵌入式的照明体系让空间变得流光溢彩。

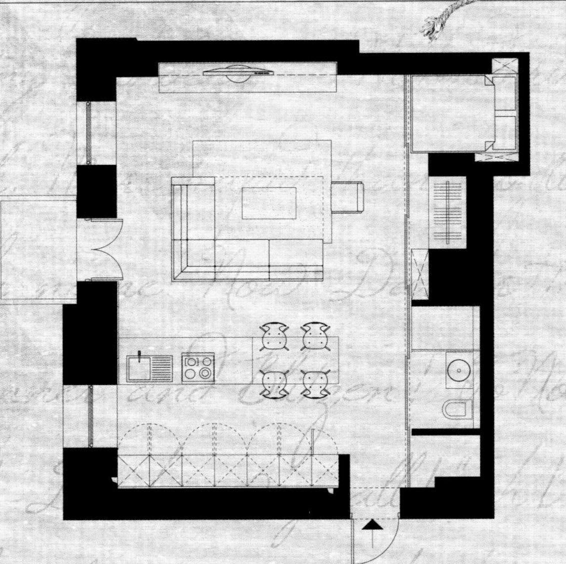

In spite of rather limited space, a young client wanted his apartment to have it all–a living room, a kitchen, a dining area, a home theatre system, a bedroom, a walk-in closet, a small library and a nice bathroom. To squeeze a full-fledged living space into 50 square meters, Makhno Workshop had to come up with a room with multiple cabinet concept. The whole apartment can be perceived as a row of cabinets hidden behind sliding glass doors, organized in a train compartment like style: a living room right at the entrance is followed by a coat closet, and a bathroom with a toilet, shower and sink. The next door leads to a library, its walls lined with bookshelves. Then comes a walk-in closet, and, finally, a tiny cozy bedroom looking more like a box with a bed inside. It is when all the doors are open that the "cabinets" reveal their true nature and purpose, giving the living room more depth and space.

An open guest area is made up of a kitchen, an elongated kitchen island with a dining area and a spacious sofa facing a home theatre system. The vividly colored modular sofa by Roche Bobois is a center of the composition, defining the rest of the interior palette. The walls are painted rich deep purple to match the dominant orange color of the sofa, echoed by Konstantin Grcic orange chairs (Magis) in the dining area.

The color with its mixed palette is not the only means to stun though, textures and surfaces are just as varied. Glossy surfaces of white custom-made cabinets (design by Sergiy Makhno) and blocks of genuine black marble of the kitchen and bathroom floor go perfectly together with natural solid wood of the flooring. Even, smooth surfaces alternate with sculptured panels (3D panels in the living room and bedroom designed by Sergiy Makhno). The image is completed with a mix of designer lights and a hidden lighting system.

随着装修材料的更新换代以及装饰设计师们思想开放度的提高，家庭装修中每年都有其流行的趋势，而当下的装修材料，如布艺、木材、石材、油漆、瓷砖和新型材料的流行趋势如下。

家纺布艺

家纺布艺的本质精髓就是要“以人为本”，不管是家纺布艺、遮阳窗饰、墙纸软包，还是家居饰品，一切家纺布艺用品都要从人性化的角度出发，因而，精细工艺、科技面料、多元功能，成为2015年家纺布艺产品市场的主流趋势，这与整体行业的大环境及消费者的消费需求息息相关。消费需求决定市场发展，随着消费者消费观念的改变以及对品质生活的追求，势必对家纺布艺用品的环保度、功能性不断提出更高要求。健康舒适的纯棉床品、低碳环保的天然纤维、经久不衰的绣花工艺，跟现代人的环保理念非常契合。

编织工艺：错综复杂的手工编织布料采用乐活的色彩，成为新颖的材料选择。布面的纹理结构成为重点，包括辫状编织、蜂窝织纹、十字绣、交织的人造皮革等，甚至可搭配民族风的图纹（如南美的梭织图案），为家饰布品带出新意。结合工艺感的手工风格，仿佛有一种与土地的连接，天然织物如羊毛、亚麻和棉布的采用，传统慢速、考究的生产方法，歌颂的是一种纯粹的原始感动。

水洗褪染：细致优雅的飘渺色调，是室内装潢的摩登尝试。床具和洗浴织物、窗帘、坐垫、帷帐、地毯等都尝试采用水彩晕染和细腻粉彩效果。扎染工艺可用于靠垫，营造休闲的嬉皮格调。蓝染效果可出现在卧室、浴室及地板。微妙的渐变褪色、手绘笔触的晕染条纹、模糊渗色的流动图案，都展现出一种浪漫的艺术风情。

怀旧之美：饱经风霜的作旧外观为家居织物带来无限灵感。提花、金箔、模糊套印、透孔织物、松散编织、天鹅绒纹理，模仿打造作旧磨损的效果。

自然原质：拒绝过度人造化的世界。2015年我们也看到了许多朴素、纯粹、未经装饰的原质色彩表现，米色或石色的天然亚光材质，亚麻、大麻、棉、帆布、羊毛等质地，凸显静谧而休闲的自然美。维可同色调变化表面的织纹与触感，如皱缩水洗、绞花纱结、编织、压花等，给家居装饰增添了一种柔和低调的天然美感。

活力霓虹：运用霓虹亮色打造活力四射的图案也是2015年的潮流趋势，以动感的色彩与新颖的材料结合，同时也影响了消费者对个性化表达的追求。至于图案，则十分多样，有印度及南美传统图纹，有绘画感的条纹笔触，也有变化性的抽象几何图形，或点状像素图案。

微光闪烁：低调优雅的金属色泽为华丽的窗帘与帷帐增添了几分精致感。相对于锃亮的刺眼高光泽，2015年比较偏好以局部光泽、幽暗油光提亮亚光面料。此外，细致优雅的虹彩光泽，铝、银、深灰或铜亮泽等深色表现以及金属箔印图案，都是值得观察的重点。

视幻几何：抽象的几何图形显得活泼又新潮，采用拼接设计搭配同色系配色或鲜亮色彩，或是以重复或重叠的图案创造视幻效果或是延伸感。比较受欢迎的几何图案包括棋盘格和条纹图案、三角形、六角蜂巢、立方体、V形锯齿、圆点等。

动物庄园：取自动物灵感的图案也表现抢眼，有灵感源自动物的写实数位印花，从兔子、鸡、鸭、猫、狗、野猪、麋鹿、松鼠、狐狸、狼到鸟类图案，各式野生或农庄动物现于床具和家居织物之上，甚至有拟人化的俏皮装扮。另外，动物皮纹印花也依旧可见，并可辅以提花织物、皮革、天鹅绒、毛巾线、扎花皮革等华丽织物为其增添纹理。

立体组织：立体表面不再局限于绗缝、刺绣、钩绣、烧花与压花等技法，还出现了贴花缝饰、剪纸折纸、浮雕雕塑、镭射加工、海绵钢眼等充满新意的立体结构。

此外，从意大利 Proposte 布艺窗帘展会上，我们亦可窥见，2015 年家纺布艺亦将往以下方面发展：一、采用先进生产技术以及纤维研究打造出新颖材料的呈现效果，通过纤维、麻制面料和高清印刷技术的组合而形成，打造出逼真、美观的图案。二、独特的图案，如迷彩、骷髅等活灵活现地呈现在面料以及窗帘上，带有一点手绘风的效果，通过亚麻、泡棉等材质来诠释一种不朽不羁的青春。三、作为绝佳的面饰材料，天鹅绒的表面展现出了奇特的色彩效果，迷幻感充斥着整体的图案与感觉。

木材

橱柜：樱桃木、枫木和桤木仍然很受欢迎，但其使用范围有所缩小；橡木、胡桃木、桦木和竹子的应用范围不断扩大；对径切白橡木的需求有所增加，尽管径切是最昂贵的生产类型，但径切板纹理通直且不易变形；白橡木与乡村风格（石灰处理或钢丝刷法技术）是受欢迎的；深色变得比白 / 亮色及中等颜色更常见。

家具：美国樱桃木是卧室、餐厅和客厅最受欢迎的木材种类；红橡木是家庭办公室家具最常用的种类。

地板：

原生态地板：即纯实木地板上油漆不上任何颜色，不加任何修饰，原汁原味的木头，这也是实木地板风格的一种流行趋势。原生态地板不仅是实木地板中又一类高端地板，在地板风格上更是一个新的引领和突破。

新实木地板：所谓“新实木”，是指三层及多层实木地板。虽然“新实木”在中国还是一个比较新鲜的概念，其实在欧美市场早已成为主流。“新实木”独特的优势还体现在花色、款式、制作工艺的丰富多样上。新实木地板的表现力丰富，在继仿古地板后，手抓纹、拉丝、烟熏、凹凸面等处理方式赋予了新实木地板不同的表现形式，丰富了室内装修。

三层实木地板：近年来，随着家居主流消费群和消费观念的转变，三层实木地板已经逐步成为市场主流产品，尤其是 2014 年，三层实木地板在行业迅速崛起，需求量倍增，在这个趋势下，2015 年将成为三层实木地板的市场爆发年。从色彩上看，优雅的灰色、温暖的棕咖色和天然的木本色，将成为三层实木地板在 2015 年春夏的主打流行色。从风格上看，清新、简约的北欧风，将继续是三层实木地板的风格主线。从产品上看，“品质、时尚、健康”将成为三层实木地板产品的三大关键词，成为 80 后、90 后等家居主流消费群除了设计和风格外，最关注的产品核心。

油漆

低碳环保概念的普及，使得人们对油漆的要求越来越高，时下涂料多色彩、个性化、环保化和多功能化发展趋势也越来越明显。环保、健康已经成为未来涂料发展的主色调，以乳胶漆为代表的水性涂料是目前最流行的环保涂料。涂料产品技术未来的突破点应该是向全自然方向发展，而且消费者更加追求新感官主义与品牌体验。

瓷砖

根据由中国陶瓷总部和ICC瓷砖联合主办的《2015国际瓷砖趋势发布——意大利博洛尼亚建材展报告暨瓷砖创新设计对话》显示，2015年国际瓷砖共有10大流行趋势，具体如下：

金属风格（Metallic Style）：2006—2008年期间，金属风格曾经是最受欢迎的风格，多年后，这种风格又卷土重来。现在，这种风格表现得更干净、优雅和精致。它们没有使用昂贵和不稳定的金属釉料，而是使用图形设计和其他材料来表现。

水泥混搭Cotto（Cement mixes with Cotto）：水泥与Cotto混搭，是意大利博洛尼亚建材展很多企业展示的一种新的产品风格。水泥砖现在开始运用Cotto风格的元素和色彩。这种砖最流行的尺寸是80cm×80cm，最受欢迎的颜色包括米色、赤褐色、米灰色、暗灰色及不同色调的棕色。

繁纹石材（Drama Stone）：意大利博洛尼亚建材展展出的石材有非常多的变化，每一个面都不一样，都有各自的特征。其中，主要质面是亚面或者柔光亚面，主要颜色是灰色和白色。

繁纹大理石（Drama Marbles）：繁纹大理石是永不过时的产品风格类型，同时是被最多企业

石材

随着时代的进步和发展，各类产品都是按照整个时代的特征和消费者的各类需求进行生产，作为建筑材料的石材也不例外。尽管石材是一种让人感觉比较死板的材料，但其依旧也有属于自己的时尚潮流感。

定制石材依然是一大主流

在个性张扬的年代，按需定制已经逐渐被现代人所接受，以80后、90后为主的消费群体，也逐渐开始追求按照自己的个性偏好来定制石材。随着消费者对个性化产品的需求不断加大，家居定制行业发展迅速，从石材到整体家居产品，定制化产品和专业化的服务已渐渐得到消费者的认可。据分析，未来的石材市场，能满足个性化设计的定制石材依然还是一大主流，经销商如能找到坚持走这条道路的企业品牌进行投资，售卖的产品无疑也将会得到消费者的青睐。

简约石材成为市场的新宠

在家庭装修中，舒适度与文化感逐渐成为人们越来越看重的要素。而简约风格的石材在更大程度上刚好符合了现代人的这些追求，特别是既时尚又奢华的简欧风格，美观又实用，正成为目前石材市场的新宠。

环保石材继续成流行趋势

如今，人们对家居健康越来越重视，消费者在装修时也越来越关注家居健康，尤其是老人和小孩居室的装修，对安全健康的要求更为严苛。因此，与装修最大关联的甲醛释放量已经成为人们衡量家居环保的一个重要指标。

据介绍，为了给消费者提供健康保障，石材可以使用的种类比木材种类丰富。如果能提供品质优异、环保标准合格甚至达到更高层次的石材产品，必然也会获得众多消费者的青睐，这也就保障了产品销量的增长。环保石材板材的选用在未来将会更为火热，并将继续成为石材的一大流行趋势。

反复生产的产品类型。这次意大利博洛尼亚建材展上，繁纹大理石主要的质面是亚面和柔光亚面，以及不同的凹凸和质感，主要的颜色有灰色和白色，也有一直流行的米色。

木纹（Wood）：木纹仍然是意大利博洛尼亚建材展上非常流行的风格，即使已经推介了很多年，它还在不断生产。不同的是这次展示了更大的尺寸，最流行的板材尺寸在1.8m（长）×20cm（宽）。木纹风格同样也更加优雅、干净和独特，最流行的色调是在灰色和米灰色之间。

砖块风格瓷砖（Brick Style Tiles）：砖块风格瓷砖绝对带来了不一样的声音并成为一种趋势。这种砖块规格和样式也出现在很多展示亮面墙砖的品牌上，色彩很多，最常见的还是白色。

超大规格（Grand Sizes）：毫无疑问，这次博洛尼亚展上，很多的品牌都有大规格的产品，而且看起来它们还会继续增长。

六边形（Hexagon）：这次的博洛尼亚展上，六边形的瓷砖随处可见！六边形出现在多种不同的规格上，小到马赛克规格，大到60cm、80cm，非常流行，就像大家相约好了一样。这形状流行的其中一个原因是技术上掌握了模具压制以及喷墨印刷花片图案的方法。

水泥风的花片（Cementine Decos）：水泥风格的花片毫无疑问是这次最多见的风格，它主要是和水泥-Cotto类的产品一起使用。各种颜色和各种风格随处可见，从黑白到五颜六色，或者大地色，甚至带彩陶风格的蓝色。

压花配件（Embossed Decos）：虽然没有水泥风那样无处不在，这种风格同样非常流行。在这种花片风格上，你可以开始看到墨水应用的进步，通过墨水在装饰性的花片表面上打造体积和厚度，也可以看到金属墨水、真正的CMYK墨水、下陷墨水和白色墨水的应用。

新型材料广泛应用

随着新型建筑材料、装饰材料的快速发展，未来的家居将变得妙不可言。如中空玻璃的广泛使用，会使家居在节能、隔音、照明上跨越一大步。比重小、强度高的塑钢材料，使家居隔断变得灵活，用此材料制作的推拉折叠式隔断，将更多地使用在厨房与餐厅、卧室与客厅之间。如用有机透明玻璃制作的卷帘隔断，用在厨房或餐厅或客厅之间，既可通风，又可方便两边的人谈话交流。

清爽度假，简约时尚

Fresh, Simple yet Fashion on Vacation

项目名称：汉普斯顿度假别墅
设计公司：坎皮思·普拉特设计
设计师：普拉特

Project Name: Hamptons Villa
Design Company: Campion Platt Design
Designer: Campion Platt

汉普斯顿位于长岛东部，未受污染的纯净海岸线由南环绕至东，是纽约上东区及好莱坞的名流们钟爱的度假胜地，甚至被誉为“曼哈顿的后花园”。然而尽管如此，这里依旧像一个原始的乡村，有自然纯朴的农场、树林和小动物，是都市人远离喧嚣、调养身心的热门休憩地。

本案是一幢位于长岛的度假别墅，屋主一家希望能赶在盛夏之前住进去避暑。于是，设计师普拉特只有两个月的时间来完成包括起居室、餐厅、家庭影院、7间卧室以及两层地下室在内的全部设计工作。对此，普拉特直言：“工期短，预算紧，是一个巨大的挑战！”

普拉特希望度假别墅能够完全带给人放松、舒适的感受，因此，他摒弃了刻板的风格限定，采用现代简约的设计手法、自然的材质以及纯净淡雅的色调，既达到了预想中的居住效果，又营造出一种浪漫的假日气氛。整个空间中很少有夸张的装饰，随处可见的是不同图案的手工地毯、造型简单的各式吊灯以及轻薄的纱质窗帘。家具也大都采用天然的木材和舒适的布艺。然而，这些丰富的元素并没有造成纷乱的视觉感受，反而被设计师处理得和谐统一、相映成趣。例如，在色彩明亮的小卧室中，都铺有纯色无花纹的地毯，而在深色系的大卧室中，则换成了大幅流线型花纹的地毯，细腻的设计在简单与繁复之间自然切换。

白色，是最简单最纯净的颜色，却也是设计中最难挖掘的颜色。白色的空间好像一张白纸，几乎可以包容任何色彩，但同时它又不可以被任意挥霍，以至于破坏了原本的美感。本案中，每个空间的四周都是干净纯粹的白，尤其是天花和墙面，设计师用不同形状的装饰线条使它们变得丰富起来。餐厅正上方是一个开阔的圆形吊顶，卧室的天花是不规则的几何造型，走廊两边整面的网格状护墙板则与富有线条感的楼梯相呼应。除此之外，质朴的原木桌椅、纹理优美的大理石吧台以及阳光房里藤草编织的家具，都散发着一种闲适的乡村气息。汉普斯顿灿烂的阳光透过窗户洒进房间，长长短短的光线在室内简洁的陈设间交错、变幻，恰到好处地衬托出白色空间浪漫无邪的优雅气质。

别墅中少有璀璨夺目或细碎繁琐的装饰，但艺术画作却频频出现，无论是在房间、走廊，还是墙壁、桌台上，总会有不同风格的绘画引人

驻足欣赏。绘画的精妙之处在于单一的形式下可以传递出各种各样的意韵：阳光房里辽阔优美的风景画将室外的风景延伸到室内，令人心旷神怡；惟妙惟肖的人物水彩令房间充盈着温馨的生活气息；为数最多的抽象画则成为装饰物和艺术品的完美结合，正如俄罗斯著名画家康定斯基所说，“抽象画是形象和色彩的一种组合和安排”，各种艺术画作在不同的位置上呈现出一种“漫不经心”的设计感，将整个房间装点得韵味十足。

Perched on the east of Long Island, Hamptons enjoys a geographical location on the coast that goes around from the south to the east, a place free of any population that wins popularity among the Upper East of New York and the Hollywood celebrities and honored as a back garden of Manhattan. However, here makes a great surprise by being kept a rural countryside, where the farmland and the forest offer a refuge for urban people to relax their body, refresh their mind and rejuvenate their heart.

A resort villa it is intended for summer holidays during midsummer. The limit of budget and the period of two months pose a great challenge to the designer to finish all sections, including the living room, the dining room, the home theater and 7 bedrooms as well as two floor basements.

Here is destined for relaxation and ease by abandoning stereotyped style but more employing simpler design approaches, natural materials and elegant hues to achieve the expected effect and shape a romance holiday experience. Without exaggerated decoration, a place it is that witness the appearance of hand-made carpet everywhere, simple chandeliers and gauze curtains. Most of the furniture is of wood and fabric. What's more surprising is that the abundance of elements keeps away of any distraction for vison, and instead, they fit in very well with harmony, fun and interest, e.g. all the bring small bedrooms are covered with carpet that is none of decorative pattern, while the large bedroom that is in dark color and embellished with carpet of stream-lined pattern. So the stress on detail has been done by switching from the simple to the complicated.

As the simplest and purest color, white is one whose potential can by no means be realized. A white space is like a piece of white paper, able to tolerate any others but deserving meticulous effort not to spoil its inborn aesthetics. As for this project, around each space there is the color of white, particularly on the ceiling and the wall, which are enriched with lines of differentiated shapes. The dining room has a broad circular suspended dome, compared with that of the bedroom, which is of irregular geometric modeling. Walls along the corridor is fixed with latticed chair rail echoing with the stairs with a strong line sense. Additionally, whether the table and chair of log, or the marble counter bar or the furniture of rattan and grass, oozes leisure exclusive to countryside. through windows, sunlight cast light long or short on the furnishings and accessories, setting off the romance, innocence and elegance to a very point.

Decoration hardly goes over gorgeous, or very detailed and complicated, but artistic paintings comes so frequently, available in rooms, aisles and onto walls and tables. No matter where they are, they are nothing but appealing and imposing, making a monotonous form to convey all kinds of connotation. The introduction of the landscape into the interior is largely due to the picture painting in the glass room, pleasant, relaxed and happy. The figures of the watercolor painting is so vivid to fill warm into the space. The paintings in their abstract sense at the largest number complete an impeccable combination of decorative item and art pieces, just like Kandinsky, a famous Russian artist said, "The abstract picture means fusion and arrangement of portrait and colors". Though put in different places at random, these art pieces are boasting the appealing taste with their strong sense of design.

EYE TO EYE

SARGENT
VAN GOGH

收藏光阴的故事

A Carrier to Gather Time Story

项目名称：安地瓜别墅
项目位置：英国伦敦
设计公司：伦敦 MPD
设计师：毛里西奥
建筑公司：安德鲁事务所
摄影：杰克摄影提供

Project Name: Antigua House
Location: London, UK
Design Company: MPD London
Designer: Maurizio Pellizzoni
Architecture Company: Andrew Jones Associates
Photography: Jake Fitzjones Photography

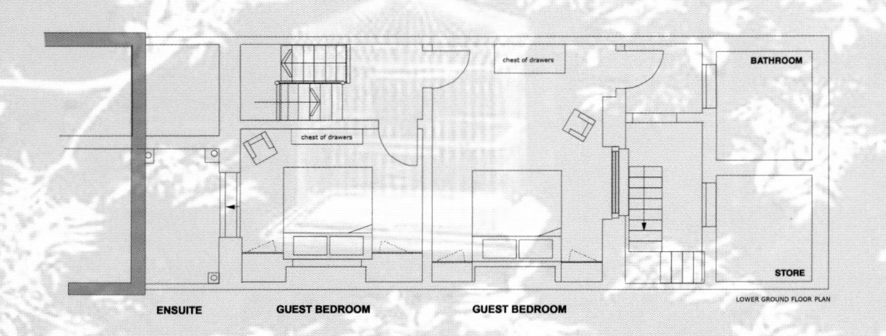

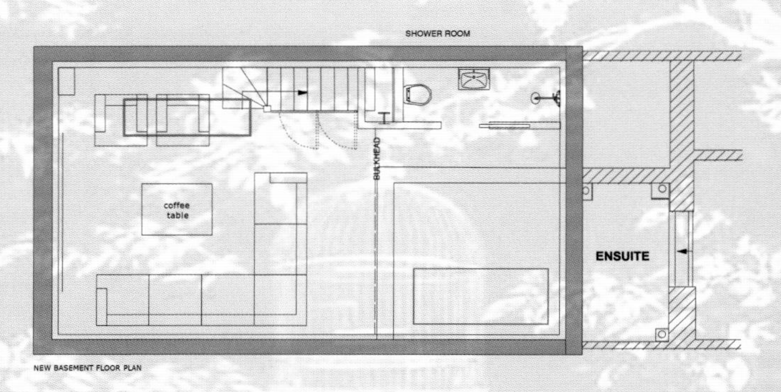

MORITZ
gadine - 6000 ft.
MORITZ
nationale Pferderennen

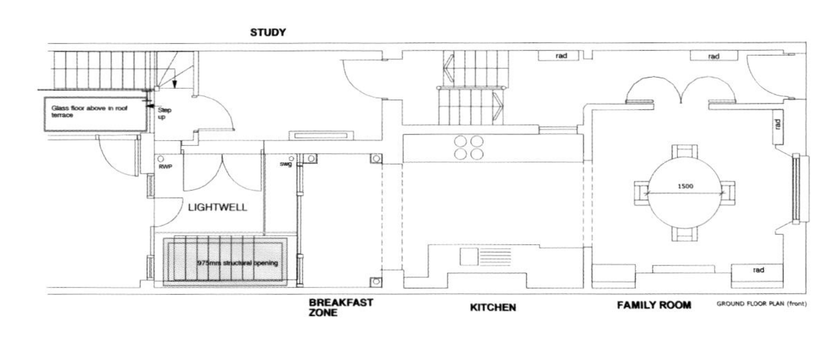

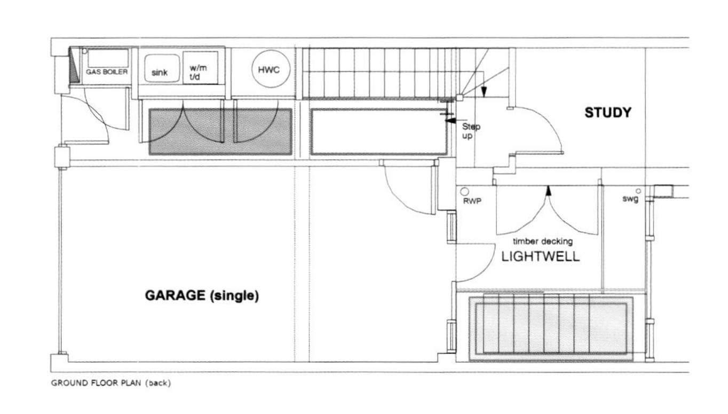

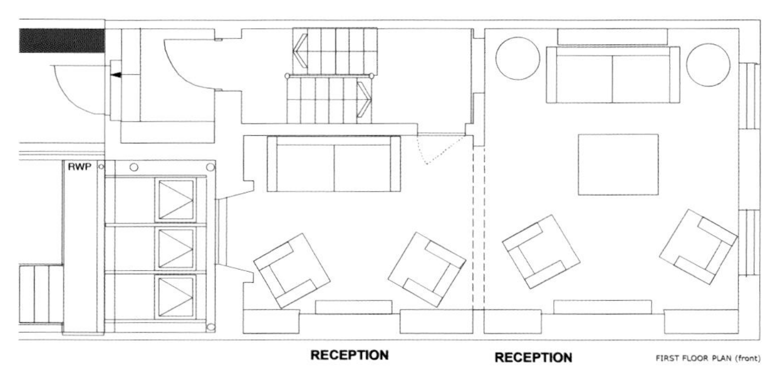

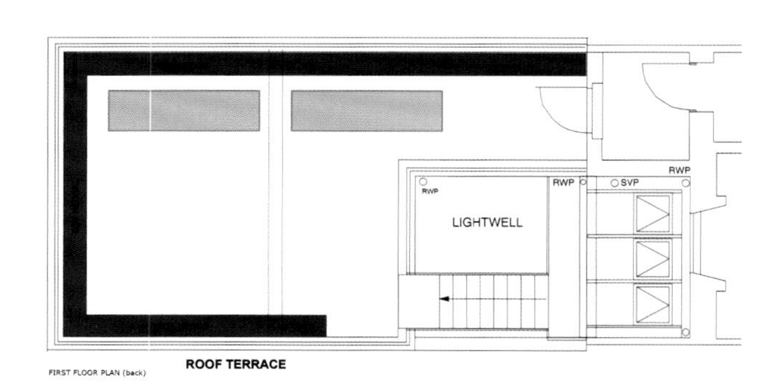

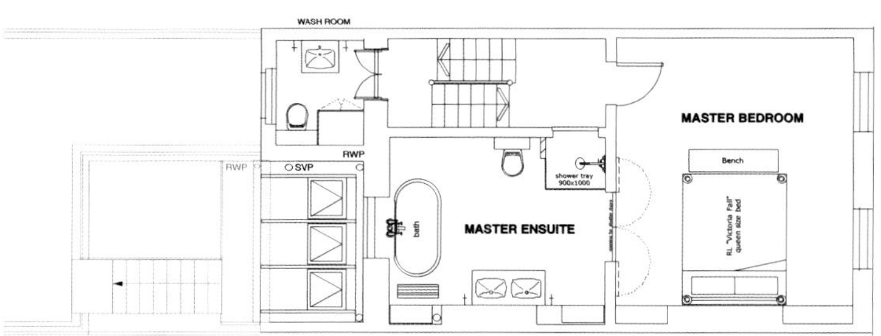

伦敦市中心的一处绿洲，宁静、祥和，便是“安地瓜别墅”所在。切尔西地区特有的文艺风华闪耀着殖民时代的光彩。富丽堂皇的木地板、自然的麻质用材、棕垫，奶白、褐灰、巧克力色的设色板齐心打造着本案空间。旧时的建筑，经过重新的平面设计，并于顶楼新添了卧室、浴室尽情满足着家人的需要。重新开挖的地下室，设计了一流的家庭影院、小小的桑拿房、蒸汽房，以及两个带有独立出入口的客卧。室外的阳台增添了透明的天窗，阳光借此倾泻而下进入下面的多功能花园书房及影院。但凡空间，于本案无不尽其所有。空间潜力得到充分挖掘的同时，亦彰显其使用之弹性。奢华的质感，富有创意的艺品，精心考究的铺陈，为家居奉献极致的舒适与别致。

ROGER MOORE
JAMES BOND 007
SOLO PER I TUOI OCCHI

LIVING IN STYLE LONDON

Antigua House is an oasis of tranquility in the heart of London. Located in stylish Chelsea it oozes old style colonial charm with rich polished wood floors, and airy rooms tastefully furnished with natural linens, coir matting and a subtle cream and taupe color scheme with accents of chocolate brown. The period property underwent major structural work to achieve a layout that suited family living, including the addition of an extra bedroom and bathroom on the top floor, and an excavated basement which accommodates a state of the art cinema room, a small sauna and steam room and two further bedrooms for guests with their own optional entrance. The outside terrace was redesigned with skylight windows that introduce daylight to the gallery room below which doubles up as a garden room/ study and through-way to the cinema. All space within the house is maximized to its full potential with reading/working areas incorporated where space is available to give this house truly relaxed and flexible accommodation. MPD finished the project with a rich layer of luxurious styling adding and selecting original artworks and accessories to create a home of optimum comfort and personality.

开启艺术之门，品味生活魅力

Door to Art, Experience to Art

项目名称：宁波财富中心示范单位
设计公司：本则创意（柏舍励创专属机构）
用材：海浪石灰、黑金砂石、夹纱玻璃、不锈钢、透光树板
面积：500 m²

Project Name: Model Unit of Fortune Center
Design Company: Percept Space
Materials: Lime, Dinas, Gauze Glass, Dinas, Stainless Steel, Translucent Wood Plate
Area: 500 m^2

财富中心地处甬江东岸，与宁波老外滩和美术馆隔江相望，作为宁波市提升战略重点工程建设项目，建筑面积达 14 万平方米，建筑高度 188 米，是宁波市建筑群中单体规模最大、高度最高的建筑。

项目根据财富中心的建筑外形，空间形态为扇形，其中圆弧部分是最佳的观景位置，设计师根据项目空间特色，划分功能区域，期望客户获得良好的生活体验。并且针对不同的空间装饰，选择最合适的照明以获得装饰上特有的情感升华，勾勒出一幅主次清晰、层次丰富、光线明亮的照明意境图。整体空间色调沉稳大气，通过大面积发光，利用天花板和墙身的造型，精准的灯光照射，形成强烈的色彩对比，渲染了空间环境的尊贵，传递出一份厚重的高贵。会客室天花中央的弧形不锈钢造型与主墙的拉丝不锈钢相呼应，在灯光的照射下熠熠生辉

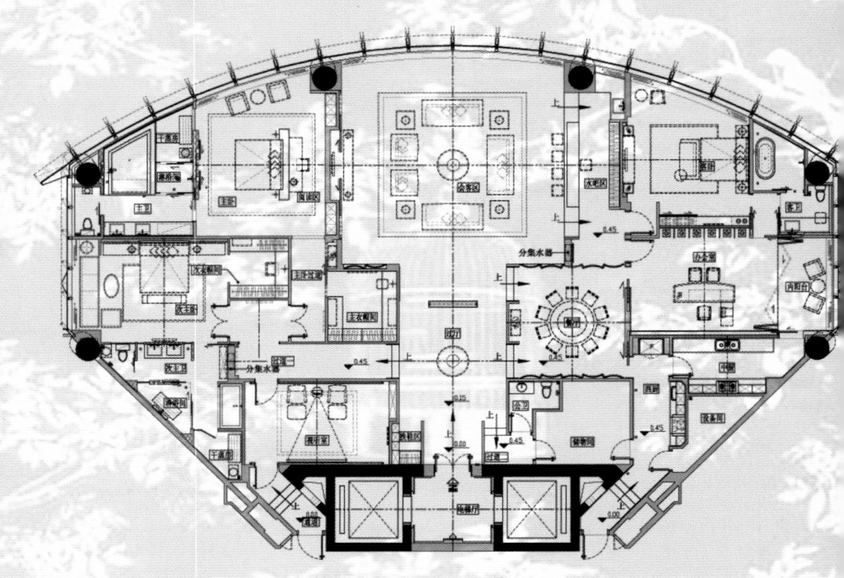

卧室以夹纱玻璃贯穿整个空间，镶嵌上鱼形的金属装饰，线条的流动赋予空间灵动性，形成极富艺术魅力的互动空间。

On east bank of Rong River, Fortune Center as a part of Municipal Key Projects to Enhance Strategies, covers an area of 140,000 square meters, measuring 188 meters and making the largest and highest building in Ningbo City.

The physical form of sector is used to allow for optimal life, particularly the arched section as the best viewing place. With functional division, the project has different decorations for different sections, each of which is fixed with the proper lighting to achieve higher emotional level and outline a prospect with layers clear and rich and abundant light. In the sharp and intended contrast of color, the spatial dignity is rendered, when conveying a historical dignity. The arched steel middle of the ceiling in the parlor echoes with the draw-bench steel of the main wall, sparking and glistering with light cast on. The bedroom is carried out with gauze window, inlaid with metal bar. The linear flowing offers the space rhyme, making an interactive space with a strong sense of art.

流动天花造型，与窗外山水呼应

Ceiling Flowing to Echo with Landscape

项目名称：画府
设计公司：城市室内装修设计有限公司
设计师：陈连武

Project Name: Line of Landscape
Design Company: Chains Interiors
Designer: Chen Lianwu

本案坐落在拥有L形落地景观窗的顶楼，依山傍水的天然格局，成为本案设计的出发点。
引用窗外的山河曲形线条，延展出室内的天花板造型。让户内与户外的氛围，能有线性的连接与融合，同时也削减了原有过多横梁的视觉障碍。墙面实木拼板与反射明镜材料的使用，创造了动态的虚实空间感，也让原有单向的景观与采光，在室内有了新的观赏角度与乐趣。以跨区域性的混搭选配，让原始有机与工业几何的强烈对比，带出了既随性又休闲，属于屋主的独特性格与样貌。

A penthouse this project makes with L-shaped floor to ceiling window that accepts the natural view on a gegoraphical locaiton near mountain and by rive that has been taken as the inspiring source.
The ceiling has an undulating profile as it projects the reversed ground topography to allow for linearity internal and external while offseting the visual barrieer by girders and beams. Materials wrapping the wall like wainscot of solid wood and reflective mirror offer a spatarial experinece both real and virtual. The nonreversing daylighting is thus endowed with new perspecitve and fun. The mashup across sections brings a sharp but intended contrast out of the industrial geomtry, embodying personal taste and easy and casual attitude for life.

细腻质感，呈现美式风情

Texture Delicate to Present American Style

项目名称：格子公寓
设计师：尤里
摄影师：安德鲁
面积：130.5 m^2

Project Name: A Checked Apartment
Designer: Yuri Zimenko
Photographer: Andrey Avdeenko
Area: 130.5 m^2

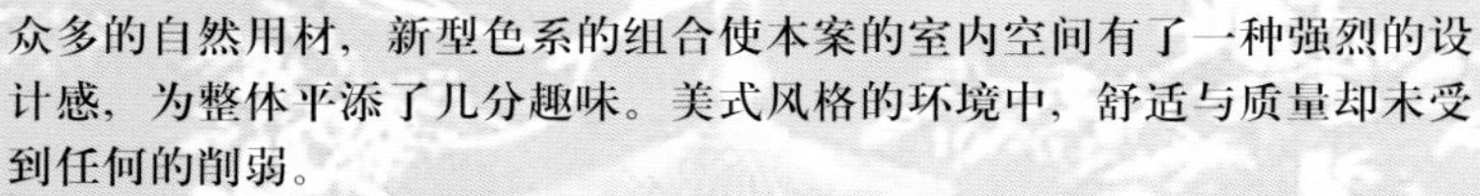

众多的自然用材，新型色系的组合使本案的室内空间有了一种强烈的设计感，为整体平添了几分趣味。美式风格的环境中，舒适与质量却未受到任何的削弱。

厨房在延续客厅非对称设计的同时，强化着地板的格子形状。餐厅与客厅之间用来界定的方格图案因为招贴画的出现而变得更为突出。地板则是富有对比的几何形状。工业美学的符号，如灯盏等，彰显着整个空间的特色。

借助于组合家具的使用，多角的客厅空间得到了架构。厨房、客厅区域除了大大的圆餐桌，还有沙发、扶手椅、咖啡桌。窗户的装饰依然为美式风格的百叶窗所代替，不再是飘窗常用的经典窗帘。

卧室更是一个放松的天下。宽大的桌，简约风格的家具，琥珀的色彩，温柔的灯光，感性地抚动着空间中一层层的触觉。

儿童房可谓是整个空间里最为宽敞的区域。设计不但考虑到学前儿童的兴趣需要，更考虑到未来孩子的发展，使设计真正做到了可持续性。

时尚的设计纹理，不同风格的方格演绎簇拥于同一空间，是设计的手法，也是本案的个性。

A variety of natural materials, original color composition and, of course, striking interior decorator's assets make the interior interesting and saturate - however, did not in prejudice of functional quality and comfort. Much americanly - which is also reflected in the style of the apartment.

Kitchen block matched in the asymmetric area of the guest space and additionally accentuated with chequered design of floor. Chess motive is supported by a poster in the partition wall between the dining and living room, and a floor carpet with contrasting geometric design. Open for viewing iron batteries and lamps in the industrial aesthetics –are the characteristic signs of loft-style.

Complex multangular of guest space is structured with the help of furniture groups. To the zone of kitchen-dining room belongs a round dining table and a sitting area is formed with sofa, armchair and coffee table. An interesting rhythm

creates a window decoration: wooden blinds (typical line in American style) are replaced by classical curtains in the bay window.

Bedroom - uncompromising space for recreation. King size bed, minimum furnishings, amber coloring and soft light, and also - sensual, carefully selected range of tactile sensations.

Children's room - the most spacious rooms in the private part of the apartment. Here not only taken into account the interests of the little fellow-preschool, but also the room has the potential of changes–according to the growth and new phases of life of its young master.

Smart decorator's vein - the connection in the same space of several types of "checked". Firstly, it is a good stylistic way to "collect" the interior, and secondly–to add individuality and style.

木色空间，营造暖怡居家氛围

Wood Space to Create Household Warm and Pleasure

项目名称：灰胡桃
项目位置：乌克兰基辅
设计师：尤里
摄影师：安德鲁
面积：200 m²

Project Name: Grey and Walnut
Location: Kyiv, Ukraine
Designer: Yuri Zimenko
Photographer: Andrey Avdeenko
Area: 200 m²

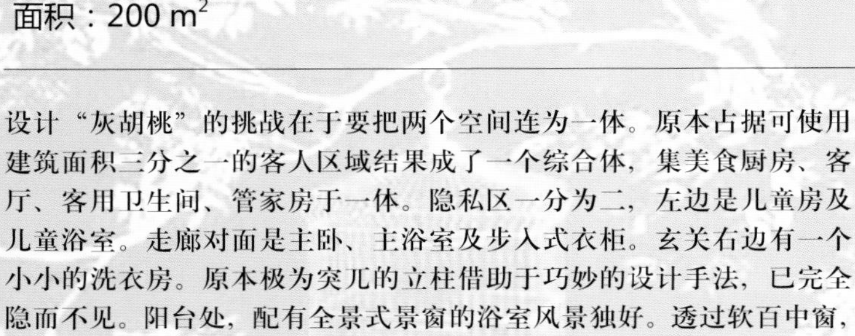

设计“灰胡桃”的挑战在于要把两个空间连为一体。原本占据可使用建筑面积三分之一的客人区域结果成了一个综合体，集美食厨房、客厅、客用卫生间、管家房于一体。隐私区一分为二，左边是儿童房及儿童浴室。走廊对面是主卧、主浴室及步入式衣柜。玄关右边有一个小小的洗衣房。原本极为突兀的立柱借助于巧妙的设计手法，已完全隐而不见。阳台处，配有全景式景窗的浴室风景独好。透过软百中窗，即便是洗浴空间，太阳也可洒下一地的斑斓。

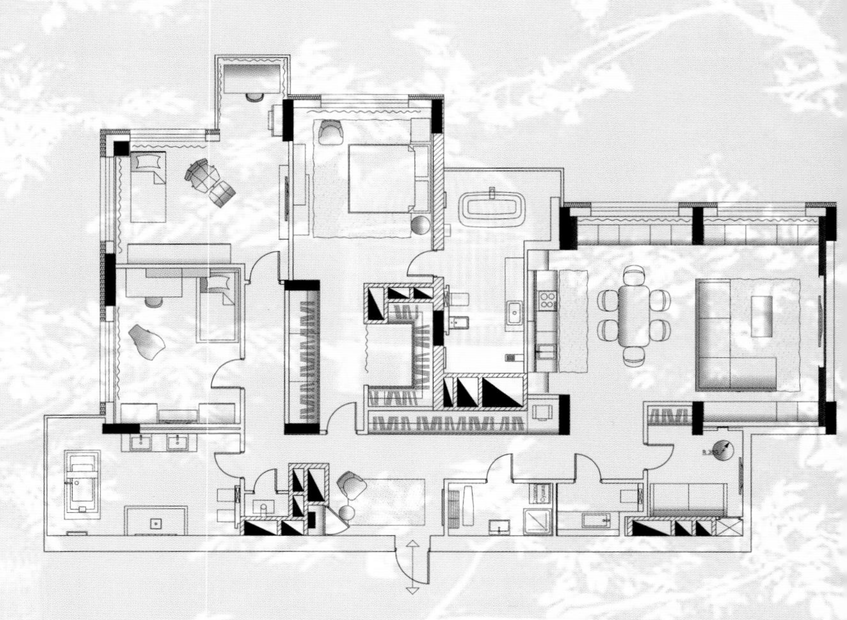

The main planning challenge was to combine the two apartments into one. Consequently, the guest area, which occupies about a third part of the total useful floor area, is composed of a combined dining kitchen – drawing room, guest bathroom and a housekeeper room. The rest space is a private area, which is figuratively dimidiated: on the left are rooms of the boys and children's bathroom, across the corridor are: master bedroom, owner's bathroom and walk-in closet. On the right of the entrance room is a small laundry room. Herewith, such a matter, as the plenty of structural columns, the designer has met absolutely in a skillful way: now they are completely invisible. On the site of the balconies sprouted vast bathrooms with panoramic windows: sunniness and extent of overlook regulated by venetian blind.

内里图画

Interior Painting

项目位置：乌克兰基辅
设计师：尤里
摄影师：安德鲁
面积：200 m^2

Location: Kyiv, Ukraine
Designer: Yuri Zimenko
Photographer: Andrey Avdeenko
Area: 200 m^2

但凡空间设计，总需要特别处理。本案空间以色调作为画笔，绘画出悬挂式的操作台。客厅以铅灰色、蓝色作为主调。窗户的帷幕延续着同样的色调，质感独好。棕色、赭色、深蓝色不时闪现其中。设计师专门设计的地毯，严格恪守着几何形状，串联起原本相互独立的色调。白俄罗斯著名画家的作品运用于其中，升华着空间的个性，设计的完整度大大得到了提升。其他区域同样运用着“色度”运算的法则。卧室以红色作为主打，并以赭色、黄色、棕色作为衬托。客厅中，地毯色彩多样，由设计师专门制作。画布般的手法如为本案专门量身打造。设计对色调的处理，同对质感的处理如出一辙。纹理、木质、金属乌光发亮。地毯纤维胶制作，顺着某一个特殊的视角，色彩如同遁形一般。窗帘也是如此。所有空间色调、风格相互统一，让空间的物体有了一种更加敦实的感觉，和谐地融于一体。

A space design required a special approach, and the first author's decision of the project was the introduction of color. And it materialized: firstly, in the form of hanging consoles. They were chosen for the living room in a compound color of plumbeous-bluish, the same color was picked for velvet curtains on the windows. Wit dusty blue took up the ball gray, brown, ocher and deep

blue. A carpet, created by the sketches of designer of the project, has united all into a single color composition, as well as "documented" strict geometry of space. The final important flourish was painting of Belarusian artists Sergey Grinevitch and Alexandr Nekrashevich, which gave individuality and completeness to the interior. This algorithm of space "colouration" worked out perfectly in other areas. In the bedroom, the red color lays down the rules of game; other colors "jig to its tune" - ocher, yellow, brown. As in the living room, a chosen range of colors captures the carpet, created by designer and attention, again, focuses on canvases - as if they were specially created for this project. There is nothing that emphasizes the designer's managing with color as his own dealing with textures. In this case, it was counted on a dull luster - textile, wood and metal. A viscose, of which the carpets are made, endows color the effect of escape: it is seen from a certain angle, then it disappears again. Such kind is shown on the velvet on curtains. The ideal job of designer with color and stylistic unity of all rooms makes the object very solid and astonishing harmonious.

低调奢华，永恒的贵族精神

Luxury Reserved, Noble Spirit Eternal

项目名称：南屿翡丽湾样板间
设计公司：深圳张起铭室内设计有限公司
设计师：张起铭
参与设计：詹远望、魏宗全
用材：石材、玫瑰金、墙布、木饰面
面积：82 m²

Project Name: Beauty Bay Show Flat
Design Company: Zhang Qiming Interior Design
Designer: Zhang Qiming
Participant: Zhan Yuanwang, Wei Zongquan
Materials: Marble, Rose Gold, Wallpaper, Veneer
Area: 82 m²

承载着新时代的贵族精神，生活空间以低调奢华的方式表现出低调而高贵、精致而有序的生活方式。空间棕色和金色的搭配，也是打造低奢的重要手法，棕色的大气稳重，加以金色的精致洗练，让整个居所充满都市新贵的品位情结！

Beauty Bay Show Flat is a carrier to accommodate the noble spirit of the new era, where the life space is flown out in a reserved approach to real life here is low-key but noble, delicate and orderly. Of brown and gold as pivotal methods to build up the constrained extravagance, the brown is of grandeur and calm, and the gold is succinct and refined to overspread the complex of urban upstart.

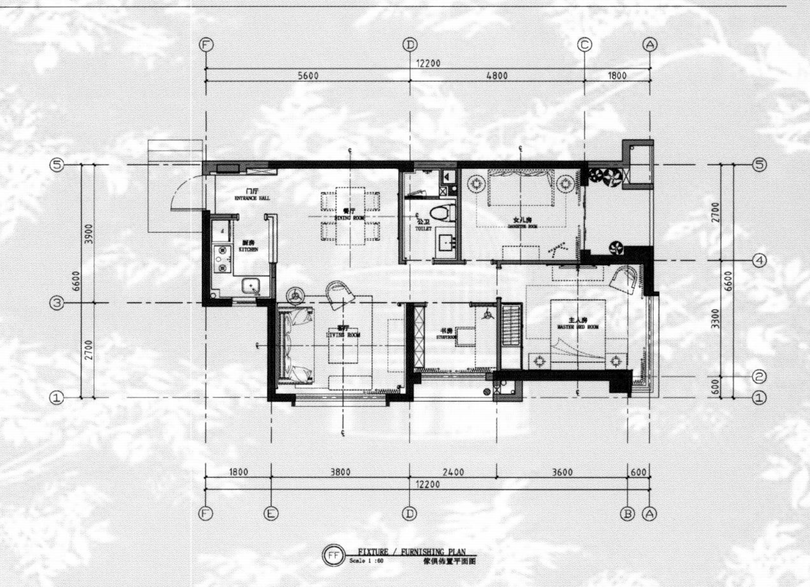

FIXTURE / FURNISHING PLAN
Scale 1 : 60
家俱布置平面图

悬空壁炉，隔而不断的空间气场

Suspended Fireplace to Partition and Link the Aura

项目名称：荷兰别墅
摄影师：雷内

Project Name: Private Residence
Photographer: Rene van Dongen

"荷兰别墅"富有纪念意义的同时，也成就了一个时髦的典范。木材、皮革塑造的空间给人一种宁静的感觉，回家的平和尽在其中。大大的生活区飞架着一个螺旋楼梯，俨然来自于城堡。整体看如同一个木作织锦毯缓缓地伸展进室内。自行设计的壁炉如同悬于空中，自然地划分着空间。伴随着楼梯产生一种戏剧般的感觉，如同立于空间中的纪念碑。楼梯上空的弯曲墙以对称的形式分成两半。

烹饪区如何做到实用而美丽是设计师长期职业经验的运用。烟熏鹅卵石铺就的浅色吧台给人一种酒吧的感觉。任人浮想联翩，但却想到与烹饪有关的任何事项。兴致来时，定制的区域有主人需要的任何器具。最新款式的美诺家具，刨冰机等给人的感觉除了豪华还是豪华。酒窖里另有酒吧，与知己小酌一杯，不会受到外界任何的干扰。

顺着楼梯便可到达主卧与浴室。浴室里陈列着令人赞叹的梳妆台。卫生间桉树木材装饰。皮质的把手强调着室内及整体轻柔的感觉。正如自然本身，因为宁静与光效的组合，在此的每一时刻都与枯燥无缘。

A recent example of Eco Chic can be seen in the monumental villa in the Netherland. The serene atmosphere of materials like wood and leather give a peaceful feeling of coming home. This big living is dominated by a stairway like we know from castles. As a soft tapestry in wood it unfolds itself into the room. In the kitchen area a splendid bar in rock crystal becomes a splendid eye catcher as it is lighted. In the living itself Kolenik designed a special fire that seems to float and which divides the room in a very natural way. The theatrical feeling increases with the stairs, now as a central monument in the room, with bending walls that high above split up in two symmetrical parts. The wooden steps make a very good contrast with the dark tiled floor in leather look. That makes a natural link with the rich leather furniture that invites to sit down. The fireplace finished by Dofine in a crusted finish seems to float. It breaks the space with a look through from a cozy second sofa.

With a long experience in design of professional kitchens Robert Kolenik knew exactly how the cooking area could be practical and beautiful. There the beautiful lighted bar in smoked pebbles offers the intimacy of a pub without an association with cooking. If the cook is ready for it he or she finds everything in this custom made area. The newest appliances of Miele, an ice crusher, a wine storage and all the luxury to feel guest in your own house. In the cellar this estate did get a second bar where visitors can get a drink in privacy.

Taking the beautiful stairs brings us to the master bedroom and bathroom with stunning make up table. In a former cupboard het integrated the toilet in eucalyptus wood. The door handles in leather underline the soft feeling of the room and of the whole surrounding. Like in nature itself the combination of serene and special effects of lighting makes the owners every day happy as here they never have a dull moment.

阔逸空间，木色相连

Wood to Bridge Spatial Broadness and Leisure

项目名称：上东联排
建筑公司：超链接
摄影师：斯科特

Project Name: Upper East Side Townhouse
Architecture Company: Hyperlink (http://zzzcarpentry.com, mailto:ninodantonio@verizon.net)
Photographer: Scott Frances

本案为翻新项目，业主希望能有现代化的欧式艺术设计。客厅的铅条玻璃是旧时的特色。曾经的镶嵌板也得以恢复，只不过经过了绝缘等更为现代化的处理，铜质的包衣外边又涂上了清水木色。餐厅、花园门、铜结构板、定制的铜板及两层的古铜色立面因此融为一体。公共空间如厨房、餐厅、客厅、图书室位于一楼和二楼。三楼有五个卧室、一个家庭办公室。而健身房、酒窖、机械房全部设置于新加的空间内。正中的天窗下，飞架着一个崭新的开放式楼梯。

For this gut-renovation project, the clients requested an updated European Arts & Crafts design. One original element was retained – the leaded glass Living Room windows. The panels were restored, laminated, insulated, and placed in the new sashes, which are clear-finished wood inside and bronze-clad at the exterior. They are integrated into a two story bronze facade, along with the dining room / garden doors, bronze spandrel panels and custom bronze gates. Family living areas, including the kitchen, dining room, living room and library are on the ground and first floor, with five bedrooms and a home-office on the top three floors, while a gym, wine cellar, and mechanical room were moved into the newly deepened cellar. A new, more open staircase, in the center of the house under a large skylight filters light into the center of the townhouse.

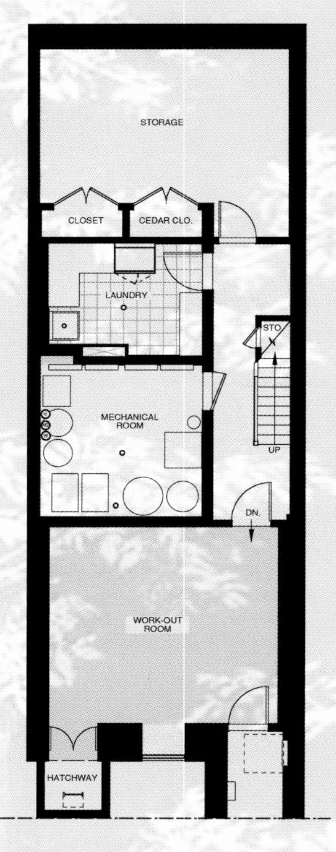

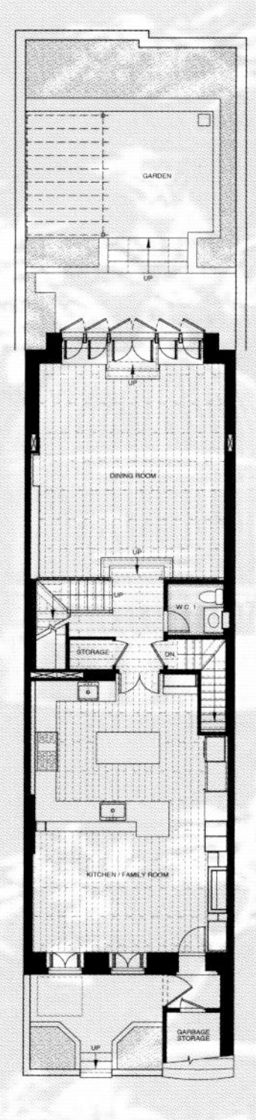

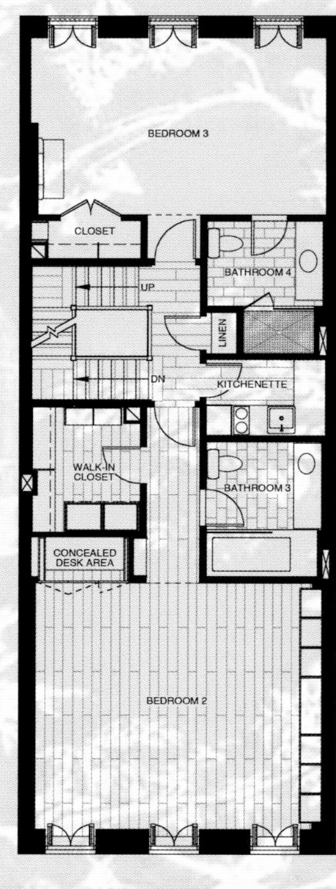

法式浪漫，佐着春光用餐

French Romance with Dining in Spring Scenery

项目名称：第五大道公寓
设计公司：巴里建筑设计
设计师：巴里
摄影师：迈克尔
面积：186 m^2

Project Name: Fifth Avenue Apartment
Design Company: Goralnick Architecture and Design
Designer: Barry Goralnick
Photographer: Michael Dunne
Area: 186 m^2

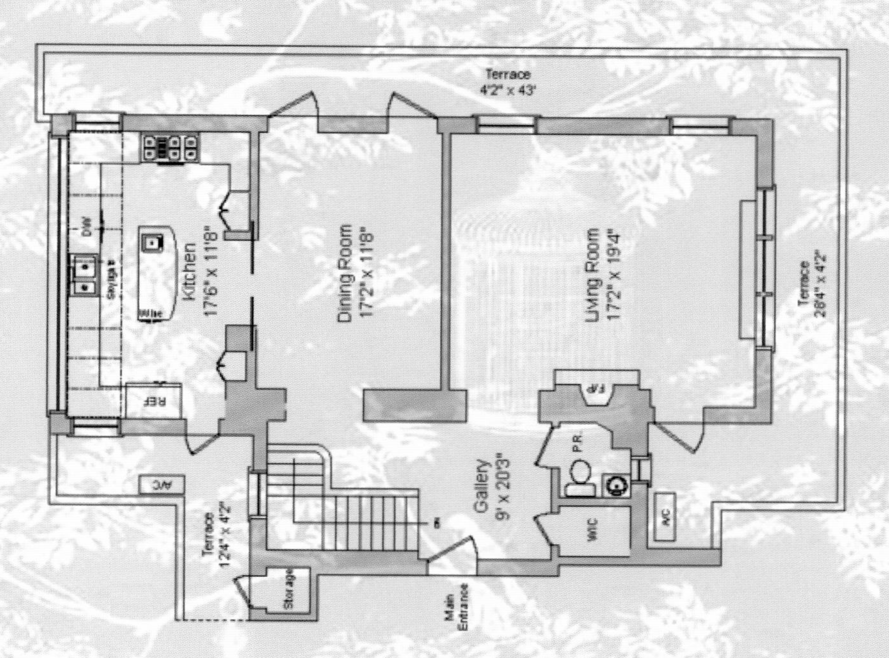
Terrace
4'2" x 43'
Kitchen
17'6" x 11'8"
DW
Wine
REF
Dining Room
17'2" x 11'8"
Living Room
17'2" x 19'4"
Terrace
28'4" x 4'2"
F/P
P.R.
Gallery
9' x 20'3"
W/C
A/C
A/C
Terrace
12'4" x 4'2"
Storage
Main
Entrance

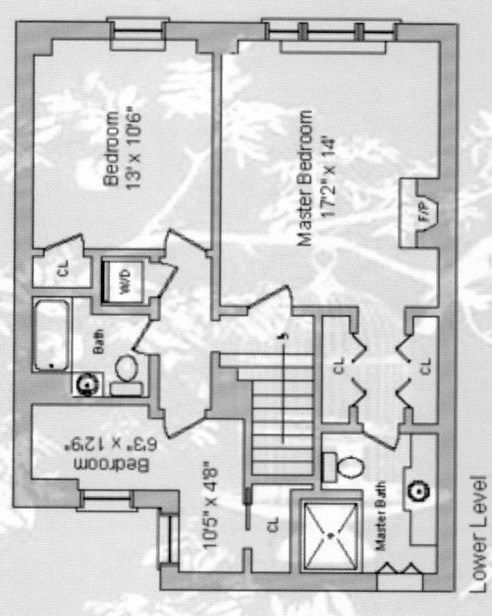
Bedroom
13' x 10'6"
Master Bedroom
17'2" x 14'
F/P
CL
W/D
Bath
CL
CL
Bedroom
6'3" x 12'9"
10'5" x 4'8"
CL
Master Bath
Lower Level

"第五大道"的业主来自英国。业主希望能营造于传统的家居空间中独觅美食的感受。原本两栋独立的奢华空间，新添的木件、门窗、中央空调、楼梯、卫生间等构成整个空间布局。厨房绿色环保，旧时的卫生间、壁橱、环绕形阳台，如今成了自然天光普照下的一个极佳的享受美食之处。

A family moving from England whose passions include gourmet food wanted a traditional look for their Fifth Avenue apartment. The gut job included duplexing two separate apartments, all new woodwork, doors, windows, central air-conditioning, stairs, and bathrooms. A green-housed kitchen created from a bathroom, closet, and a portion of the wrap-around terrace provides an ideal space with abundant natural light for the cooking enthusiast.

简约空间，收纳法式浪漫

Simple to Collect French Romance

项目名称：中央公园西
设计公司：巴里建筑设计
设计师：巴里
摄影师：桑切斯
面积：279 m²

Project Name: Central Park West
Design Company: Goralnick Architecture and Design
Designer: Barry Goralnick
Photographer: Hector Sanchez
Area: 279 m²

“中央公园西”的业主为知名演员。夫妇俩购得的此处公寓，规格宏伟。该公寓建筑风格古典，而室内家具铺陈风格折中，其中有很多法式精品。本案得以成就是与业主及很多美国知名艺术家合作的结果。来自世界各地的演艺界、学术界的朋友，但凡入内，总能与业主共享快乐、舒适与放松。

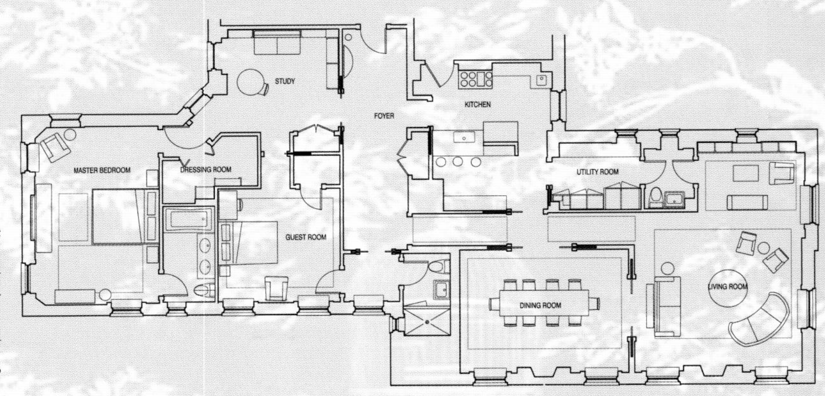

An acclaimed actor and his wife purchased this large apartment on Central Park West. The architecture is classical in style and the furnishings are eclectic, mixed with many fine French pieces from the Forties. We closely work with the clients to coordinate the art collection that showcases many great American artists including McNeil, Eisner, Freed, Gahagan, Loew, deGroot, and Pace. This home serves as an elegant and relaxed place for entertaining family and friends from the world of theater and academia.

随着世界信息的无界交融，大众视野不断开拓和审美品位的快速提高，家居装饰逐渐从功能性消费，向功能到审美、到艺术的方向发展，“轻装修，重装饰”的观念已经成为家庭装修的主流观念。与此同时，家居装饰品亦打破了传统的装修行业界限，将工艺品、纺织品、收藏品、灯具、花艺、植物等进行重新组合，形成一个新的理念。家居饰品的种类也从以前的单一型转变为现在的多样化。因此，家居装饰品要想获得设计师的青睐，不仅产品本身要具有时代性，而且产品所呈现的艺术性还要起到引领时尚潮流的作用。

在如今的消费市场中已经被定位为潮流的产物的家居饰品，与家具跟时尚挂钩不同，其更新的周期要比家具短很多，这就要求家居饰品的设计者要有足够敏锐的触觉去接收随时都处于更新状态的潮流信息。那么，未来的家居饰品又有着怎样的流行趋势呢？

一、家居饰品进入轻时代

告别了繁复精美，家居和饰品正应和着极简、环保、健康等主流关键词，进入了一个全新的时代——轻时代。无论是设计师还是制造商，都不约而同地转向简单的款式和绿色的材质，也许并不能使外观尽善尽美，甚至一些购买者的传统习惯尚未接受新材质的运用，目前并不具备主导市场的能力。但由于它的生命力与良好前景，必将成为未来的主导。

1. 极简艺术

随着无印良品、Jil Sander 等时尚的潮流所推，极简的文化已越来越成为现代都市人对家居、饰品的追求。纯粹的几何图形、简洁的线条、透明的材质、轻巧的外形，不同层次的色彩，赋予这种表现形式更丰富的情感要素。设计，即是将严谨和令人思考的元素巧妙结合，创造极简艺术。时髦的“禁锢”风格，将空间从厚重的物体中释放出来，展现出最纯粹的美的本质。

你可以看到棱角分明的几何图案，一片目无杂物的清新留白，或是整洁有序的排列组合，在纷扰忙碌之后，能收拾好自己的心情，走进一片满目清凉的境地，正好是给心情放假的良方。沉闷的华贵色彩已渐渐被明快的纯色拼贴所取代。在饰品中，尽管伊丽莎白时期的经典款式与钻石这类贵重宝石依旧炙手可热，但水晶、布艺等材料已成为后起之秀，在如今的饰品市场反而占有了很大的比重。施华洛世奇 Swarovski 的璀璨证实了人工宝石也能绽放完美光芒。

2. 绿色环保

在绿色环保备受追捧的今天，人们越来越呼唤纯天然的绿色家饰，自然清新的装修风格已成为时尚，这也使得当今的家居饰品趋向清新、自然、古朴，甚至是原始的野味。采用各种新颖、别致、具有时代感和装饰效果的材料装扮居室时，环保的饰物成为了消费者更高的追求。

此外，在居室内种养花草，已经逐渐成为老百姓美化居室环境的一项重要手段。当然种植绿色植物也是修身养性的一种好方法，如今也渐渐成为一种潮流。而植物与静物的结合使家居与饰品变得鲜活富有生命力。一种回归自然，返璞归真的心态，使人们在庸扰中得以解脱。试想，当你居住在充满植株的室内，佩戴生长着花草芽苗的饰品，你的心情是否会比在拥挤杂乱的房间看着硬冷的金属放松一百倍？

二、家居饰品设计融古汇今

现在是一个古典与现代结合的时代，古典家居样式的现代演绎已经是现在装修设计的亮点，能满足人们追求现代时尚的同时也喜欢古典韵味的要求。精致的装饰、绚丽的材质，不仅更多样化、人性化，在色彩设计上也突破了中式古典的格调，加上了现代的感觉，色彩的多样性、年轻化更受欢迎。在装饰设计上，古典的刺绣、雕花等与现代科技相结合，赋予了室内空间更多的文化底蕴和现代情感，比如市场上流行的陶瓷台灯等。

三、家居饰品实用化

家居饰品除了美观和别致，还增加了趣味性以外，更强调了实用性。美观独特的外形依旧存在，但是，美而有用、美而有趣，成为了附加的必备前提。未来家居饰品将实现实用化潮流趋势，做到每件家居饰品作为家居用品的实用价值，同时实现实用产品的艺术化，根据产品风格设计符合产品风格唯一性与关联性的家居饰品。

四、时尚类家居饰品变换加快

时尚类家居饰品包括一些造型可爱的小用品等，由于这些家居饰品设计不够经典，所以变换较快。采取以不变应万变的对策，挖掘永不过时又显档次的家居用品。软木大行其道，裸色系百搭风格，适合与任何色彩与材质搭配，让家中不断变换的其他软装因为有了软木用品的点缀，张扬出主人的独到眼光及生活品质。

五、感性家居饰品让家有故事发生

感性化、个性化的设计，当然就有感性的解读，以前形容家居饰品常用的是“时尚”和“养眼”等词汇，而未来越来越多的设计师开始在家居饰品上通过图案、色彩或是感性的文字编织着一个又一个的甜美故事，让家有故事发生，感受家居另一种温馨和温情。

六、家居饰品品牌化

家居饰品的品牌一直是国内家居饰品制造商的一个软肋，随着一些国际家居饰品品牌陆续进入国内家居市场，作为一种时尚的消费，家居饰品已经不再是价格上的竞争，而上升到了品牌和设计上的竞争，模仿已经渐渐失去了市场。走进各大商场的家居饰品销售部，在老面孔之外你看到的更多的是时尚品牌的推出。也就是说，很多人已经开始习惯购买各类时尚品牌。

事实上，很多时尚品牌也开始出现时尚家居饰品，像 ESPRIT 品牌之前以服装为主要经营目标，但近年来也推出了自己的家居用品。国内的家居饰品卖场在做大、做专后要注重培育品牌，只有这样才能与国际家居饰品品牌抗衡，从而赢得更多的市场。

家居饰品的流行趋势在保持其基本原则：个性、提升自我的同时，更要不断变换着方向、各种风格、材质、色彩轮番登场，从不同的角度、不同的形式，美化现代时尚家居。对家居饰品的流行趋势的充分理解，以及对其阶段性流行元素的充分把握是始终站在流行舞台中央的保证。因而，不断关注家居饰品的流行趋势，无论是对热爱生活的人们，还是对家居饰品从业者来说，都是必要的。

收藏艺术，也收藏激情燃烧的岁月

To Collect both Art and Time of Passion

项目名称：福熙城

Project Name: Foch Downton

本案设计的目标是希望能赋予空间情感，创造一种与众不同的美感。富有弹性化的设计可以让空间随着时间的流逝而实现“型变”。

橡木地板，蛋壳色调的墙面塑造的调色板衬托着主人心仪的家具铺陈与其他艺品收藏。室外的古罗马遗迹如若不加以利用，真是“暴殄天物”。中性调色板的适时出现，引入了室外的风景。一切色调所有与室内各得其所，各就其位，产生一种碰撞。

从拱门般的镜面控制台到定制的亮光吧台，无论是质感、用材还是造型均刻意突显对比。吧台的边角是锯齿形状，富有新鲜感。虽说个性十足，但一切还是以家居用户为本。

各房间色调、图饰自然搭配。主客厅白色基调，给人一种放松、欢迎的气氛。明亮色彩的图案地毯衬托着众多的艺品，空间的色彩有了点缀与跳跃。其中一幅绘画还是达利的大作。

主卧，Versace 金属质感、松绿色卧床融于整个空间，并彰显着其中的特色。抽象的拼贴画金色镜框，恰好成了床头的背景。该画由 Manuela Crotti 所作，手笔不凡。

In this house, the aim was to evoke emotions, to create something different and beautiful, and to allow for enough flexibility to let the home evolve in style with time.

A soothing palette of solid oak wood floors and eggshell color walls provides the perfect backdrop for the owner's collection of interior pieces and art. The views of the roman ruins outside are phenomenal and it would have been criminal to overpower them. The predominantly neutral palette allows for the outsides to come in; and also allows all the color in the house to own its place and make an impact.

From the mirrored console that references archways to the bespoke high-gloss bar with its jagged little edges, Vick has masterfully contrasted finishes, materials and geometries to create a space that is very individualistic and yet remains true to its inhabitant.

Colors and patterns where chosen to match the moods for each room. The main living area is white for a relaxing, inviting ambience. A bright patterned carpet and some vibrant art pieces including a Salvador Dali, punctuate the space delicately with color.

In the master bedroom, a Versace metallic turquoise bed blends in perfectly with the whole room decor; and a magnificent collage art work with golden frame, by Manuela Crotti, make the bed head board.

Coca-Cola

كوكا كولا
Coca-Cola
Coca-Cola

ANTHONY ROBBINS AWAKEN THE GIANT WITHIN

B.B.M.
SIZE

阁楼上的好莱坞之梦

Hollywood Dream in Attic

项目名称：顶层公寓
项目位置：墨西哥圣·路易斯
设计公司：S&K 室内设计
设计师：雪莉、卡蒂
摄影：马大摄影提供

Project Name: Penthouse
Location: St. Louis, MO.
Design Company: S&K Interiors
Designer: Shirley Strom, Katie Marvin
Photography: Matthew Harrer Photography

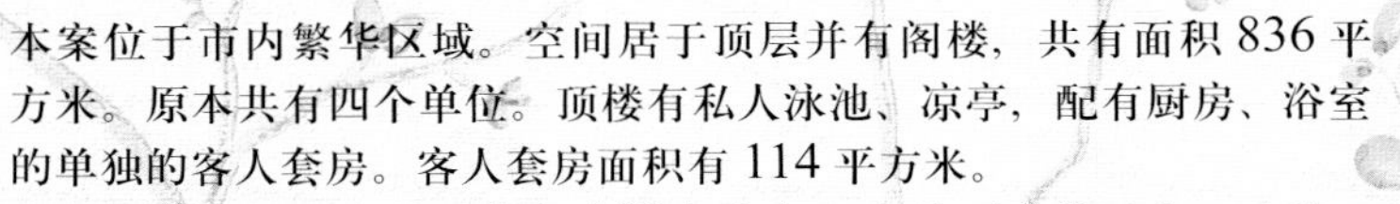

本案位于市内繁华区域。空间居于顶层并有阁楼，共有面积 836 平方米。原本共有四个单位。顶楼有私人泳池、凉亭，配有厨房、浴室的单独的客人套房。客人套房面积有 114 平方米。

空间所在建筑距今差不多有 100 年之久，参差不齐的天花、砖墙、水泥墙、屋顶及其他各元素保持完好。老式的电梯间如今成了主卧的沐浴间。

修缮后的新旧空间在此汇集，用材考究，面貌焕然一新折中风格，现代、紧跟潮流彰显着好莱坞的魅力。

This project is an innovative gut renovation in downtown St. Louis, MO. Our client decided to purchase the entire top floor of a downtown building along with the rooftop. The new penthouse is almost 836 square meters; it combined four previously separate apartments into one fabulous open floor plan. The rooftop contains a private pool, lanai, and separate guest bedroom suite along with kitchen and bath, which is another 114 square meters.

The renovation was in a century-old rehabbed manufacturing building, leveraging the 15'-18' high ceilings and existing brick and cement walls, the rooftop and other original building components. One feature is an old elevator shaft which serves now as the new master bath shower.

Almost the entire apartment was gutted and replaced with new materials. S&K Interiors came up with an eclectic design style mixed with modern, contemporary, transitional, and a touch of Hollywood glam resulting in a fascinating mix of old meets new.

TOM FORD
ARAKI

TOM FORD
ARAKI

Campbell's
SOUP

NEVER HIDE

CHANEL

KARL
WHO?
POP
Meier
HADID

And I said I Love You!
CONTEMPORARY ARCHITECTS
Madame
Monsieur
LOVE ART LONDON

I LOVE COCO CHANEL

I LOVE COCO CHANEL

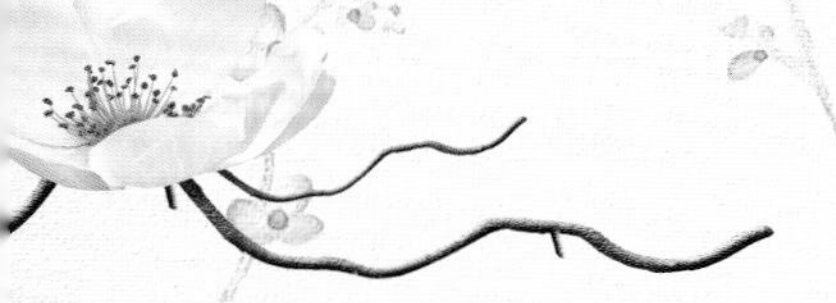

北欧风情，以细节成就品质

European Style: With Details to Make Quality

项目名称：布宜诺斯艾丽斯公寓
设计公司：AMC 建筑设计
设计师：安东尼奥、马里奥
摄影师：马克
面积：225 m^2

Project Name: Buenos Aires Apartment
Design Company: AMC – Architecture and Interiors
Designer: Antonio Ferreira Junior, Mario Celso Bernardes
Photographer: Marco Antonio
Area: 225 m^2

本案公寓风格古典，重新设计后的空间与建筑原有线条迥然相异。空间木器、护壁板、地板，甚至一些门户等多为建筑师抢救性所得。厨房、客厅面积上得到了扩大，与阳台合二为一，功用彰显现代化特征。墙面涂饰茴香色、淡草绿色、鸭蛋黄、油蓝色，色泽不一。

The apartment had a more classic style, different from the original lines of the building. "I brought inside the art deco facade elements present in", explains the architect who rescued woodworks, baseboards and trims the season. Interference predicted for him also restored floor, and some doors, modernized the use of the property with the expansion of the kitchen and the living room, which was coupled to the balcony, and even dyed the walls with shades of the season as fennel, pistachio, yolk and oil blue.

05
QUARTA
MAIO

Minimalism

橙黄橘绿溢彩居

Dwelling in Orange, Yellow and Green

项目名称：圣保罗公寓
设计公司：AMC 建筑设计
设计师：安东尼奥、马里奥
摄影师：佩德罗
面积：353 m²

Project Name: Itacolomi Apartment
Design Company: AMC – Architecture and Interiors
Designer: Antonio Ferreira Junior, Mario Celso Bernardes
Photographer: Pedro Setubal
Area: 353 m²

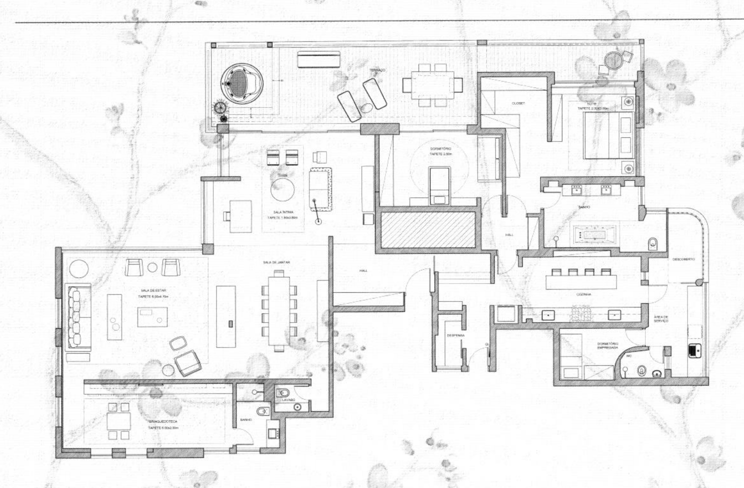

现代风格的空间里，铺陈着来自里约热内卢、圣保罗、纽约的古董器物。借助于设计师的妙手，整个空间涂饰着各种经典色调。家庭影院中的灰色板块中点缀着红黄。餐厅里一个巨大的铁制造型极为夺目。同样的着色用于大厅里的红木条凳。金色大理石的地板让空间多了一种优雅的气氛。家庭影院因此优雅地与露台融为一体。露台的墙面涂抹着 SPA 的绿，蓝绿色的质感因为木地板、绿植而得到了升华。露台的墙体界定着内外空间，但一条走廊的设置使内外之间得到了沟通。

In a contemporary style which brings out vintage pieces mined in Rio de Janeiro, Sao Paulo and New York, the architects Antonio Ferreira Jr and Mario Celso Bernardes managed print your brand colors abounding in accurate measurement throughout the apartment. The speckled gray predominates home theater with red and yellow. Dividing the home dinner, a huge panel coated iron stick counts the dining room. The same coating can be seen in the lobby with adding an elegant rosewood bench. The marble floor Calacata Gold in addition to extending the space creates an elegant atmosphere where brilliance imposes the home office that is integrated to the terrace with spa-green wall with a beautiful and imposing where colors like turquoise is supported by wood floors and green plant. The wall of the wall as well as giving color and life Social to terrace separates the outside of the inner layer, but which retains the integration through a passage.

Christian Dior

Christian Dior
KATE MOSS BY MARIO TESTINO
VALENTINO GARAVANI
ROSE, C'EST PARIS

当代生活主义：睿仕魅力绽放

Contemporary Life: Farsighted and Genteel

项目名称：合肥中海滨湖公馆
设计公司：上海桂睿诗建筑设计咨询有限公司

Project Name: CSCL (Hefei) Lakefront Mansion
Design Company: G&K International Design Institution

在男人眼里，车子无疑占据着非常重要的位置，它富有激情并充满生命力，它激发了男人潜藏在心底深处的控制欲，成为男人感情生活体验中最美好的一部分。当读不懂一个男人时，我们常通过他对车的态度来识别其气质类型，男人对车的态度就是男人对生活的态度。让我们通过他的车所蕴含的设计理念，带给他一个属于它的家吧。
整个空间以米灰色和咖啡色系为主色调，而橙色和金色的加入则起到画龙点睛的效果。整体色彩体现男性的硬朗成熟、内敛沉稳又不失潇洒风趣的气度。明亮的客厅，米灰色的地板与壁面，白色与橙色的沙发座椅，前卫现代的家具和灯饰，使客厅不再单调，俨如一个时尚又悠闲的高级沙龙。餐厅大理石纹路地板搭配方形不规则灯泡，给人强烈的视觉冲击。而主卧室以褐色为主色调，给人以沉稳的感觉。
当然成功男士除了事业有成以外，他们对奢华物品、高级享受的追求依然热情十足，细节一丝不苟。他们与现行社会体制也不会构成太大的冲突。他们工作勤奋，不胡乱追求物质享受，但重视家居品位与生活质量。因此，在本案室内设计，各种奢华物品遍布，更有轮胎、机车服、摇滚音乐器材等引入其中，既满足了实用机能，又具有强烈的装饰作用。这样的设计，表现了主人以热爱生命的方式来演绎精彩睿智的人生，不是一般对名牌盲目追捧的伪雅皮或真土豪可以拿来相提并论的。

Male eyes make pretty large room for cars, as they stand for passion and life vigor and vitality when motivating the control desire hidden in the depth of heart and playing a significant part in males' emotions. When hard to read of a man, we can judge him by his attitude towards cars, which is actually his attitude towards life. And so, this project offers a man a dwelling space with life philosophy that comes along with cars.
The holistic persistently coated in beige gray and brown, the finishing point is made with the use of orange and gold. The

ROCK
BAND

whole color applied within embodies the muscle and the mature of the male, while he is equally unobtrusive, staid, natural and unrestrained. The bright parlor is really likes a sophisticated salon, where the gray flooring and walls, the white and orange sofa, and the modern furniture and light fixtures make here everything but nothing boring and dull. In the dining room, the marble-vein flooring equipped with irregular bulbs give a stronger visual impact. As for the master bedroom, it is wrapped in brown to feature staidness and steadiness.

Besides a successful career, distinguished man still exerts zeal over extravagant goods and sumptuous enjoyment when particular about details. They have no large conflict with the current society, they are diligent at work, they don't go blindly after material, but they lay more emphasis on household taste and life quality. That's why in this project all luxurious objects go with tire, locomotive wearing and rock music. That is practical, aesthetic and decorative to embody that the host live his life with love for life, incomparable with those Japy who thinks of nothing but seeks after name brand or those Tu Hao, literally meaning local tyrant in the past but now usually referring to people with fortune but without taste.

DESTINY
DESTINY
DESTINY
DESTINY

DESTINY

色彩唤醒空间、陈设打造风情

Hues to Awake Space, Furnishings to Make Style

项目名称：北京鸿坤林语墅
设计公司：LSDCASA
设计师：葛亚曦、李萍
面积：361 m²

Project Name: Forest Villa
Design Company: LSDCASA
Designer: Ge Yaxi, Li Ping
Area: 361 m²

本套别墅设计面积361平方米，是鸿坤集团在北京的又一豪宅力作，项目以“爱丽舍宫”为范本，从规划到建设，皆以皇家宫廷为参照样本精工雕琢，成就“京城新贵”最后的城市低密生活。

整个室内空间，设计师以色彩和陈设来实现不同区域的特定功能和独特韵味。首先大量运用中间色系让空间融合起来，然后点缀以小面积深紫和灰蓝的撞色。中西方元素的搭配，令整体沉稳优雅的空间具有变化。公共区域中，设计师采用视觉感强烈的图案、地毯、挂画来烘托一个区域的特定氛围。家具的选配方面，多以美式风格家私搭配现代风格的边柜、灯具等，材料在绒与麻之间还加入了金属、水晶元素，视觉感十分丰富。每个区域都有吸引人之处，并充分展现主人儒雅的个性和豪爽的热情。

With an area of 361 square meters, Forest Villa makes another mansion by its owner. With Elysee as its blueprint, the project refers to French imperial palace to present the last low density community for the newly-rising riches in Beijing.

Hues and furnishings on the whole work to partition specific functions and unique flavor. Neutral colors fuses the whole space with dark purple and gray blue in small area as well as elements of the east and west setting off the holistic grace. The public space is embellished with pattern of strong sense, like hanging carpet and pictures to highlight certain ambiance. Furnishings mainly consist of those of American style with modern side cabinets and lighting fixture. Materials of velvet and linen are added with metal and crystal to allow for a wealth of visual effect. Each section has its appealing while embodying scholarly personality and forthright passion.

一重一重打造空间焦点

Focus in Layers Layer Out of Space

项目名称：利安娜

Project Name: Leanna

设计如能突破风格、色调、质感产生一种与众不同的视野，必将是作业时的一大乐趣。波普艺术的触感与工业时髦融于本案空间，最终促成了本案。

生活白砖墙纸，并有定制的“班克斯”碳纤维所作“生活如此美好”的绘画作为修饰。如此强烈的视觉中心点，衬托着空间的乐趣、色调、动感，并与现代的混凝土、木桌子共同反映着业主的个性。

圆形玻璃艺术品上绘制的保罗·罗宾逊限量版的杰作，正对着下面特别醒目的红色绒布沙发。非常实用的皮质椅，同时可用作书籍、艺品的收藏。中央的桌子以几何的条状金属作为底座。色调柔和的几何形小方毯强调着视觉中轴，不同的元素因此紧密相连。

家庭室兼用作电视室，以同样的元素混搭，只不过方式更为精致。里面的电视柜，可谓是维克招牌之作。餐饮区，除了富有自然色调的墙面，木质的桌台、板凳，灵感源于中世纪的橱柜，同样是维克的杰作。

This is another fun project that aims to push the boundaries of design and dare to experiment with mix of styles, colors and textures to produce the unique vision and style that is Vick Vanlian. Urban, industrial chic with a touch of pop art is the mix for this project.

Dominating the living space is a white brick wall paper with a customized banksy reproduction graphite quoting: "life is beautiful". With this strong statement making the focal point, the fun, colorful and dynamic mood of the house is automatically set, and in turn reflects the owners' characters, mixed with urban industrial chic concrete and wood center tables.

A fabulous striking red velvet sofa is strategically placed under a limited edition Paul Robinson print on glass circular art piece, and in turn, contrasted with a practical leather bench, which is also used as a storage space for books and valuables. Geometric metal strips base for the center table, and geometric, pastel colored, Paul Smith rug emphasis the visual axe and helps to bring the different elements together.

The family and TV room continues the same mix of elements in a more subtle way, and hosts the signature red and grey cabinet by Vick Vanlian. The dining area, rich with natural wall nut wood table top and bench, mixed with Midcentury inspired cabinet, produced and stylized by Vick Vanlian.

L'Art du Moyen-Orient
CARS & STARS
COSMOS Giles Sparrow

施华洛世奇之美

Aesthetics of Swarovski

项目名称：荣禾·曲池东岸二期 E 户型
项目位置：陕西西安
设计公司：SCD（香港）郑树芬设计事务所
设计师：郑树芬
参与设计：杜恒、丁静
摄影师：叶景星
面积：195 m²

Project Name: Unit E, Second Phrase, East Bank of Qu Pond
Location: Xi'an, Shanxi
Design Company: SCD
Designer: Simon Chong
Participant: Amy Du, Jimmy Ding
Photographer: Ye Jingxing
Area: 195 m²

摒弃繁复厚重的古典欧式，进行简化创造，结合都市元素为年轻一代带来充满活力的现代简欧家居。新与旧、低调与奢华、优雅与随意的有机结合，碰撞出精彩的设计火花。由色块、几何图形、线条组合的抽象挂画点缀在空间各处，恰到好处地提升空间的现代设计感，各个细节凸显独特的魅力，一直蔓延到整体，赋予空间更多的雅致与奢美，完美体现了郑树芬的"雅奢"设计主张。

浅色调的绒布沙发既典雅又现代，线条简约的施华洛世奇水晶吊灯为客厅带出轻奢感，地毯和挂画以简单抽象画为主，大气的笔法和泼墨的轻松和自由感，正体现出现代都市人的向往。

印有克里姆特经典作品《吻》的陶瓷装饰，唯美的拥吻画面昭示着温馨、浪漫和富有激情的生命力，充满爱的力量，与施华洛世奇的纯洁之美相得益彰，升华客厅空间的感染力。

蓝紫色圆形地毯、湖蓝色圆形蒲公英图案餐椅，搭配璀璨的水晶吊灯，塑造出愉悦、轻松的就餐氛围。香芋紫色的三人绒布沙发与浅棕方格布面沙发相混搭，深棕色的窗帘则延伸了家庭厅的层次感，简洁的白色立柜、金色的造型摆设与黑白的几何抽象画渗透出现代都市气息。

步入主卧犹如走进浪漫的色彩世界，柠檬黄色的床品点亮整个米色空间，丝绒墨蓝色窗帘，嫩粉色桃花，欧式花纹壁纸等都令人仿佛置身于花花世界，感受到春天般的芳香与温暖，充满生机与力量。

空间也是有灵魂，有生命力的，施华洛世奇超凡品质与精湛工艺，以其散发着光彩魅力的水晶吊灯为空间注入非凡的精神象征。那隐藏在温润触感之中的闪烁水晶，伴随着设计师娴熟的手法，将爱的力量，生命激情融汇于空间的每一处，演绎出流露着施华洛世奇之美的精神魅力家居。

A space this project is that has abandoned the complicated of

THE COLLECTION

THE COLLECTION OF
SIMON CHONG DESIGN

the classical European but undertaken more simplification in combing urban elements to offer young generation modern simple European furnishings vigor and vitality. The new and the old, the reserved and the luxurious, and the elegant and the casual are organically blended to bring forward design marvel. Color lump, geometric pattern and hanging picture of line group in its abstract sense everywhere boast the modern design sense to a very point. All details with their prominent charm are available throughout, providing the space more elegance and luxury and well embodying the design intention of grace and luxury of Simon Chong.

The velvet sofa of light hue is both elegant and modern. The simple-lined Swarovski chandelier gives the living room luxury. The carpet and the hanging picture are mainly of simple, abstract pictures, whose magnificent techniques and easy and free brush-ink embody the desire of urban people.

The ceramic decoration is *Kiss*, classical works by Gustav Klimt. The kiss of aestheticism implies warm, romance and vigor and vitality of enthusiasm and passion. That is the power of love, setting off the good quality with the purity of Swarovski, consequently enhancing the appealing of the living space.

The circle carpet of bluish purple, the dining table and chair patterned with lake-blue dandelion and the chandelier shapes an enjoyable and relaxing dining atmosphere. The purple-yam velvet sofa for three people is mixed with light brown check fabric sofa. The dark brown curtain extends the layers of the family room. The simple white clothes closet, the golden modeling and the black-white geometric abstract picture confide in a strong sense of metropolis.

The master bedroom seems to have made a world of colors, where bedding of lemon yellow lights up the cream-colored space and the velvet curtain of very blue, the pink peach blossom and the European-patterned wallpaper expose you to a land of flowers, where you feel nothing but fragrance, warm, vigor and vitality.

Any space has its soul as well as its vigor and vitality. With superb quality and fine craftsmanship, Swarovski chandelier injects into the space extraordinary spiritual symbol. The glistering crystal in the warm texture beneath the skillful hands implant power of love and passion for life everywhere, accomplishing a house where to flow out the aesthetics of Swarovski.

探寻空间的旋律

Explore the Melody of Space

项目名称：北京中粮祥云 B 户型
设计公司：LSDCASA
设计师：葛亚曦、李萍
面积：500 m^2

Project Name: Forest Villa
Design Company: LSDCASA
Designer: Ge Yaxi, Li Ping
Area: 500 m^2

在葛亚曦的每个设计作品里，我们很难看出明显的风格，但事实上，他做出来的作品却往往能让人有惊喜的感觉。他常常把很多元素混搭、融合在作品中展示，以此把握空间背后的关系。在他的作品中，我们甚至能看到草间弥生装置艺术的影子，或是意大利先锋设计师的作品风格，有时候甚至加入一些批判的元素，用这样多元的手法呈现出其不意的惊喜。

暖棕、麻灰、米白构成了空间的主要色谱，内敛沉稳，这些源于自然的色感韵律令人身心放松。除此之外嫩黄、紫红、深蓝等闪烁着欢快的亮色，跳进人们的视野，它们的出现不仅带来了好心情，也与空间中的灯光交相辉映。所有包裹空间的材质、承载着视觉特质的色彩和悉心甄选的艺术品，共同勾勒出环境中的一切，令每一个空间散发着不同的能量和气场。

It is difficult to figure out the obvious difference among Ge Yaxi's works. However, his works can always bring us surprise. He usually integrates a lots of design factors into his works in order to deal with the relationships behind the space. We can usually find out the characteristics of Kusama installation art and the design styles of some distinguished Italian designers in his works. Sometimes, he adds some critical factors to his own works. His works can bring us some unexpected surprises by using multiple design factors.

The warm brown, heather grey and white are the main colors of this mansion. These natural colors make the mansion seem calm and restrained, making people feel relaxed when step into the mansion. In additions, there are some bright colors, such as bright yellow, purplish red, dark blue and some other bright colors. These bright colors come into people's eyes, bring not only good mood, but add radiance and beauty to the whole space. Everything in the mansion, including materials, colors which carry the unique visual characteristics and the carefully selected art works, outlines the whole atmosphere and makes every space have different characteristics and aura.

朋友的家

Friend's House

项目位置：基辅
设计师：马诺赫
摄影师：安德鲁
面积：150 m²

Location: Kyiv, Ukraine
Designer: Sergey Makhno
Photographer: Andrey Avdeenko
Area: 150 m²

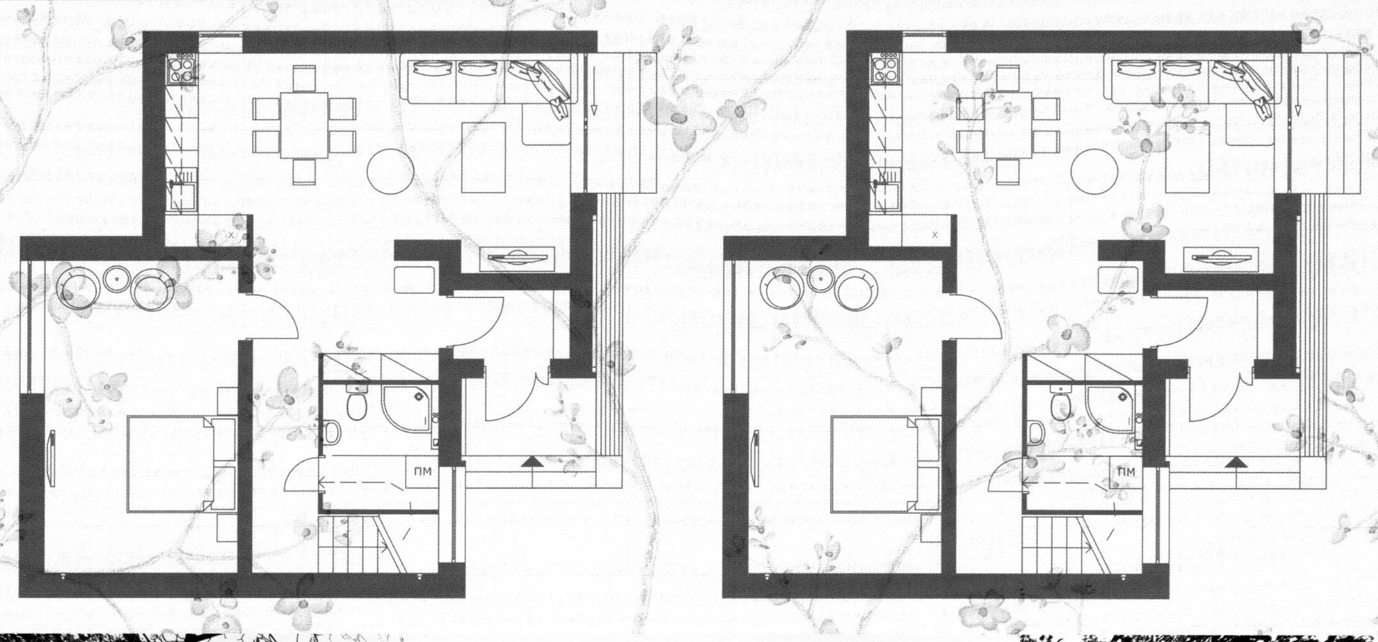

x3 score600
CIRCUS

如果本着放松休息的原则，乡村别墅并不真的一定多么高端。如果业主热衷于旅行、收藏，则空间无论如何都不能枯燥、乏味。本案业主是设计师的老朋友，很显然并不是艺术、设计的门外汉。

两层的空间占地150平方米，足以容下如厨房等各功能空间。除了一楼有一个卧室，二楼还有两个。另外，空间还有两个卫生间及几个组柜。甚至空间还有一个供喝茶、悟禅的地方。如此这般的设计，真可谓是思想信马由缰。绘画的墙面、水泥的楼梯、天花的木壁板、木地板、瓷砖地板等，一切是那么简单，但却恰如其分。空间的气氛似乎如同为其中的家具铺陈量身定做。意大利著名品牌B&B的标志性沙发、扶手椅，20世纪富有传奇色彩的Achille Castiglioni设计的落地灯。另外，书桌、控制操作台、木质的斗橱等等全部彰显大师手笔。透过大面积的全景式开窗，室内可观农庄的秀美景色及烧烤区。雕塑大师米特罗（Dmytro Grek）的杰作“迸发”（burst inside）摆放于空间，如同天成。空间内还有米特罗的其他作品与其他名家的瓷雕及艺品相互衬托。

每一个房间都摆放着独一无二的器物与别致的艺品，给人留下经历难忘的印象。茶室最出乎意料，长长的低矮瓷桌、染色玻璃花艺图案、锈铁立方架创利出一种特别的情绪——乌克兰的禅意。其实，这是本案空间关键的创意手法。

A country house doesn't really call for any sophisticated solutions if it is meant to become a perfect place for rest. But if its owners are into traveling and collecting stuff, there is no way the house will turn out boring. The owners of this house also happen to be Sergiy Makhno's old friends and obviously know a thing or two about art and design.

Two storeys with the total area of 150 square meters are quite enough to accommodate the essential: a combined kitchen/living/dining room area, three bedrooms (one on the first floor, two on the second floor), two bathrooms and some closets. Even a room for meditation and tea ceremonies, which is, strictly speaking, more of a whim, has found a perfect place here. Everything is as simple and organic as it gets: painted walls, a concrete staircase, wooden paneling of the ceiling, wooden (the second floor) and ceramic tile (the first floor) flooring. The atmosphere of the place is born out of recognizable design objects and art: the iconic sofa and armchair by B&B Italia, Vitra chairs and the 20th century's legend — Arco floor lamp by Achille Castiglioni. The interior concept is completed with Book Table (a glass table with the legs made of high voltage insulators), Up console and a wooden chest of drawers with a concrete top (all by Makhno Workshop). The kitchen set was made by the local artisans based on Sergiy Makhno's designs. Through huge panoramic windows facing the courtyard and a barbeque area, you can see bronze sculptures by Dmytro Grek that "burst inside", becoming an integral part of the interior. Some of his works have found their place in the house, living in a perfect harmony with ceramic sculptures by Sergiy Radko and Yuri Musatov, and art pieces by Vlada Ralko, Petro Bevza and Viktoria Melnychuk.

One of a kind objects and unique art pieces lend each room its very own unforgettable charm, where the tea room stuns as most unexpected. All objects found here — a long low ceremonial table, a screen with stained glass floral patterns, burlap pouffes, rusty iron cubic shelves designed by Sergiy Makhno — create that special, unmistakable mood, which is suddenly seen as the workshop's key creative method — Ukrainian Zen.

雪人先生的家

The Home of Mr. Snowman

项目名称：绿地国际城 A2 户型样板间
设计公司：成象空间设计
软装设计：成象空间设计
面积：92 m²

Project Name: A2 Show Flat, Greenland International Town
Design Company: Imaging Space Design
Upholstering Design: Imaging Space Design
Area: 92 m²

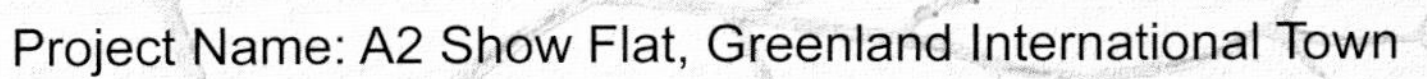

这是北欧最为典型的家居空间，四面墙体素雅净白，回归简朴，解放心灵，搭配具有设计感的家具，美丽但不复杂的装饰，让人感觉非常舒服，满足了业主喜欢极简风格，希望家中干净整洁又有韵味的愿望。客厅的布置最为典型，以素色为主色调，以蕴含春意的装饰品作点缀，让空间更灵动，更随性。大窗户让空间邂逅春日的阳光，邂逅南归的飞燕，偶尔迸发的几抹亮色让空间停留在初春的自然气息里，把家从冬日的慵懒中唤醒。嫩芽初开的枝条在碧绿花瓶里安静等待，是等待春天，等待未来？还是等待主人回家的脚步？而餐厅的白幔吊灯，则营造出若有似无的情调。

主卧素雅的白色与咖啡色呈现出情感细腻的现代风貌，背景装饰品是由柔韧的竹与生冷的铁结合而成，暗合"蒲草韧如丝，磐石无转移"的心意。每一件配饰都可以使房间呈现出别致的风景，鸵鸟先生伴着关不住的花果香气，成为家中新奇别致的焦点。儿童房跳跃的黄色和饱含生机的鲜嫩绿色，让房间充满童话色彩。底色虽素雅，但角落里大大的E.T.、随手摆放的变形金刚和相框，让空间并不死板。

每个人心中，总住着一个少年，爱手伴、爱乐高、爱缤纷的马卡龙、爱五彩的生活。

This project makes a typical dwelling place that comes across from North Europe, whose simple yet elegant style is bound to fly free heart. With a strong sense of design is beautiful but not complicated, the furniture allows for personal preference of minimalism, with which to provide the owner clearness, tidiness and connation.

The most remarkable living room is dominated with shads and decorated with ornaments that imply spring, so that the space becomes more ethereal and causal. Through the large window, the warm sunlight comes across the south-returning swallows. Some patches of bright colors keep the early spring lingering within, the holistic up to wake up from the indolent winter. Budded branches are quiet in the very green vase, right there waiting for the spring, or for the future, or for the returning steps of the families. Out of the white chandelier in the dining room, comes an emotional appeal that seems to be here but somewhere unperceivable.

Arietta

The white and the brown in the master bedroom sketch a contemporary appearance, appealing and subtle, whose backdrop is made with bamboo and iron, fitting in with the intention that the stem or leaf of cattail can be as tough as silk, while a massive rock can by no means be transferable. Each and every item boasts the unique in the space. The ostrich and the fragrance of the flower and fruit have now been a special focus. The contrasting yellow and the green in the child's room overspread the flavor of fairy tale. Against the simple and elegant hue, here is not dull or tedious at all, particularly due to the large E.T. in the corner, Transformers and photo frame that are arranged at random. There lives a teen in the depth of each heart, who has a passion for Lego, for Macaron and for multi-colored life.

清新淡雅，舒适无价

Elegant and Fresh for Priceless Comfort

项目名称：熨斗公寓
设计公司：巴里建筑设计
设计师：巴里
摄影师：桑切斯
面积：167 m²

Project Name: Flatiron Loft
Design Company: Goralnick Architecture and Design
Designer: Barry Goralnick
Photographer: Hector Sanchez
Area: 167 m²

本案的业主是位作曲家，希望家庭录音需要的清净与频频的客人来往互不干扰。于是有了本案，曾经的旧工厂华丽转身后成了私家府邸。空间除了两个卧室，还有一个录音工作室。设计集工业时代的感觉与公寓的有条不紊、隐私于一体。

A composer wanting a loft look and feel with aural separation for both home recording and frequent guests purchased a space in an old factory building converted into apartments. The loft was divided into two bedrooms and a recording studio. It is a blend of the industrial feeling of the loft and the organization and privacy of an apartment.

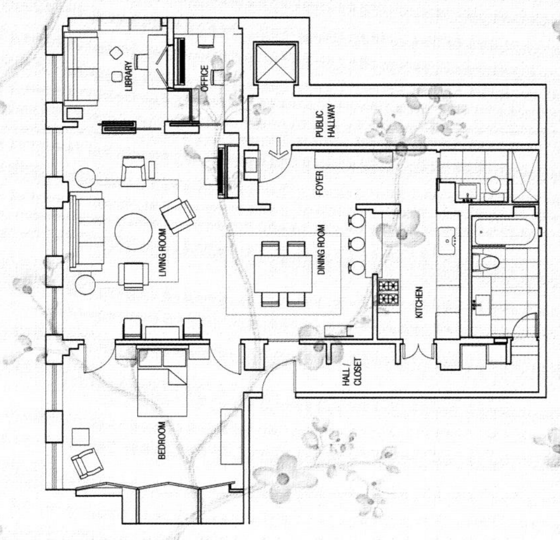

叠

Overlap

项目位置：中国台湾
设计公司：加娜设计工作室
设计师：林仁杰、陈婷亮
用材：木皮、漆、玻璃
面积：66 m²

Location: Taiwan, China
Design Company: Ganna Design Studio
Designer: Lin Renjie, Chen Tingliang
Materials: Veneer, Lacquer, Glass
Area: 66 m²

休息大厅的左面是一个L形的岛柜。柜体嵌入立柱式的设计把不同的区域紧密连接在一起。大厅右边，是80厘米宽的长凳的预留地，借此希望把大厅引入室内空间。沿着大厅左转是一个长长的桌面，集餐桌、书桌为一体。桌子右手边的设计拉伸了整个空间的视觉感。

主墙镶嵌电视，界定着主卧与客厅之间。通往主卧的动线有两条。如果右门开启，三面邻近的窗户均可采光。室内自然光线普照如同白昼。走廊、通道的照明集中于桌子、茶几。一旦灯光开启，气氛别致怡然。因为空间玩具等陈设样式多样，展示柜面呈以烟灰色，以便烘托里面的藏品。凳子浅木色，桌子棕色。三种不同的色度，借助于黑色而统一于整个空间之中。如黑的大型水晶吊灯，黑的大抱枕，不规则的地毯，黑的格子沙发。

三盏水晶吊灯以不等距形式间隔出现，目的是为了使空间具有更为生动的感官印象。抱枕的尺寸比常规的80cm × 80cm规格要大，正好装饰着客厅。不规则形状的地毯也起到同样的功能。而黑色的格子沙发则烘托着整体的气氛。

The left side of the lobby is a L-shape kitchen island. The designer deliberately extends the island to the column in order to connect different areas. The right hand of the lobby was planned to place the long bench, which extends from the lobby to the house. The width of the bench is 80 centimeters. Turn left from the lobby can see the long table which combines the dining table and study table together. On one hand, this long table extends the visual feeling of the whole space. On the other

MY GUN
PETS

MY GUN

hand, it can provide enough seats if there are some visitors.
The main wall on which inlays TV separates master bedroom and the living room. The designers arrange two traffic lines to enter the master bedroom. People can selectively open each door depending on their needs. When the right door of the main bedroom is opened, three windows close to the door can share the light source. Consequently, in daylight natural illumination can be introduced to the house. It is almost unnecessary for the indoor users to turn on the light. As for the aisle and passage, the arrangement of illumination will be focus on the table, on coffee table, on the wall and those areas the users would stay more frequently in order to create special atmosphere.
Since the colors of toys and miniatures are various, the designers use ash gray to be the ground color of display cabinet so that those collections can be highlighted. And, because the colors of display cabinet (ash gray), the bench (light wood color) and the table (brown) are in different color hues, the black color in the soft decorative assembly was adopted to integrate different areas. For example, three large chandeliers, large pillows, black cushion on the bench, irregular-shaped carpet, and the black plaid sofa are all dark colors or black.
The arrangement of three chandeliers is not equidistant so the visual display can be more vivid. The size of pillow is considerably larger than general pillow (80cm × 80cm), adorning the living room. Irregular-shaped carpet does the same function. Besides, two designers also choose the black plaid sofa which can warm the overall atmosphere in this area.

告别形式主义的铺张，告别奢靡，新一代的都市新贵们，不再满足于张扬于外的富丽堂皇，更注重与内心相呼应的高贵品质。展望2015年家具流行趋势，风格上中式、欧式、美式将继续流行，后现代受到追捧；在80后、90后等年轻一代逐步走上社会舞台，他们带来品位情趣上的差异将引领家具设计走上“简约、时尚、文化、定制、智能”的风潮。

趋势一：全球化

如今的全球化不再是一味的简单复制西方文化，而是双向融合后产生的全新风格。一方面，美国和欧洲越来越简约的设计趋势，将影响中国家具设计；另一方面，中式文化的复兴，中国设计正以全新的姿态融入世界并影响世界。此外，随着经济更加密切的融合和互联网的深度渗透，全球化带来的影响已经进入到人类生活更深的层面。不同地区文化因子由于人和信息的加速流动而被传播到以往不能达到的地方，产生了许多实验性的方向和突破。非洲风、希腊、罗马、埃及、波斯等地区的文化，甚至包括一些更加小众的文化元素混搭而产生的创意，迅速出现在各种不同的产品方向中，并拼构成全球艺术的整体。

趋势二：糅合东西方美学精华元素的混搭风格

源于2001年日本时装界的混搭，已经很难考证是谁把它引入建筑、室内装修、美食等领域。当今时代的发展从根本上改变了人们的生存方式和行为方式，加之互联网的普及发展，使文化的传播更加迅速、明朗化。欧美风格的大热和中式的复苏，使混搭——将东方时尚与西方美学传统交融——消弭界限，而又保留各自独特的表现，设计适合中国人使用的欧美家具，寻找东西方文化的平衡，在未来两年的家具行业中继续“得宠”，并进行得更加彻底和深入。又因消费人群生活方式和潜在审美需求的差异化，不断衍生变化。

此外，随着中国GDP的高速增长，APEC峰会上中国全球影响力的展现，以及在亚太地区经济合作中所发挥的重要作用，中国正受到世界的高度关注。作为中式文化载体的现代中式家具，必然也将受到国内外消费者的追捧和喜爱。

趋势三：跨界潮流

如今越来越多产品、室内、建筑、软装等行业设计师的跨界，以及大量出国攻读设计专业留学生的归国，不仅为家具行业带来更先进成熟的研发模式，而且提升了整个行业的品质。不少汽车、奢侈品品牌也都跨界进入家具行业，如Armani、Aston martin等。

趋势四：全屋、个性定制

从高端定制到全屋定制，定制已经成为未来的大趋势。全屋定制更是从传统的活动家具延展到定制衣柜、橱柜、鞋柜，以及门、楼梯、门厅、飘窗等，帮助消费者解决所有装修问题，实现拎包入住的理念，统一的风格调性获得最佳的搭配效果。此外，工业化生产既创造了丰富的物质，同时又留下了千篇一律。造型、色彩与雕刻无太大差别的家具产品，让人们难以投注欣赏的目光。个性化定制旨在打破这样的僵局，根据每个人或者每个非常小的群体实现按需定制。

趋势五：智能化

国际智能家居市场，引来互联网巨头谷歌和世界最顶尖的科技公司苹果的强势介入，智能家居也被称为下一个千亿美金规模的市场。随着物联网、云计算等战略性产业的迅速发展，小米和美的的牵手，TCL 与京东、海尔与阿里巴巴的战略合作，“工业 4.0”概念的提出，智能必将成为家具行业未来发展的大趋势，部分家具也将采用更多的现代科技手段，比如更全面监测睡眠质量的床垫、多功能的酒吧柜等，实现声、光、色、形的最佳匹配效果。

趋势六：重视环保

2014 年期间，国家颁布了多项与环保有关的家具行业法规、标准，以及《濒危野生动植物种国际贸易公约》附录内容的生效，环保已经成为家具业的大趋势。同时，消费者对“健康”概念的全新认识和关注，也让环保产业成为新消费热点。如曲美新品——万物放弃传统的板材和实木，采用更环保和可持续发展的竹钢。

2015 年，将有更多的注意力聚焦在环保领域，当然也有更多的环保绿色行动。家具本身的环保，比如有害物的控制、加工过程中使用的辅料环保性能等，都会摆上高端家具品牌们的案头。

趋势七：多功能

就目前的购房需求来说，以小户型居多。小户型的房子由于面积限制，家具太多、太大的话会导致使用面积不足，这就使得拥有第二功能的一物多用家具深受消费者追捧。由于多功能家具可以更好地利用空间，使得“人们希望家具可以拥有多重用途”已经成为一种流行趋势，并且这种需求正在以更快的速度增加。

最常见的一物多用家具就是沙发床和储物沙发、储物凳，但随着科技的发展，多功能家具的多功能已经不仅仅体现在使用功能上，还体现在新技术、新材料的融合上，已经演化为实现其他新设功能的现代家具类产品，是对家具的再设计。多功能家具的四个显着特征是：机电一体化技术的融合、计算机技术的应用、新材料的应用、精巧的可调式构造。一件家具至少具备其中一个特征才可称为多功能家具，如折叠床、升降折叠餐桌、升降柜等。

趋势八：极简设计

都市的快节奏生活，让人始终处于精神紧绷的状态中，繁冗的家具设计只会让人觉得累赘，极简主义设计的家具以其回归原点的简单造型、优雅流畅的设计感和黑白的基础色系，用最简洁的造型，获得最大的舒适感，始终站在世界家具潮流的前端。目前，消费群以 80 后、

90后为主，所以，纵观2015年米兰国际展展示的产品，都以简约为主，不管是古典家具还是现代家具，均舍弃过去繁琐的雕工、华丽的外表，以简洁、优良的材质、价值感来展现低调的奢华。

极简风格虽然简单，却不空洞，多数设计是简单的线条，不带过多的曲线，富含设计或哲学意味但不夸张；除了极简主义的代表色——黑白之外，灰色、银色、米黄等无印花无图案的原色也带来另一种低调沉稳内敛的宁静感；极简家具的材质除一般的皮质、木质之外，还有各种现代工业的新材料，如铝、碳纤维、塑料、高密度玻璃等，使家具具备防水、耐刮、轻量、透光等特点；虽然造型简单，但同样具有多功能性，如在可塑性最高的椅子部分，极简设计的椅子多有功能性，可自由调整高度、变化造型；床架可打开成为另一处置物箱；橱柜打开后收纳功能强、桌椅可拉开变宽等。

趋势九：新时尚、轻奢华

在设计圈，时尚是永恒的主题。家具追寻时尚，但又超出时尚。从服饰、汽车、箱包、大自然、各种文化中汲取新的灵感，都是时尚的表现。当然，大胆、前卫的色彩搭配，从未见过的雕刻图案等，都是时尚。与此同时，高端家具设计也正开始遵循低调奢华、轻奢华路线。它在设计中吸取了多元古典、温馨舒适、个性化、高科技等多种元素，并且着重体现与众不同、格调高雅，追求高端尊贵的艺术视觉，创造生活的顶级享受。

趋势十：情感表达

设计是有生命的，家具也是有生命的。一种有质感与生命力的家具设计，虽不能自我言语，却能满足业主对居住空间的渴望，能够为业主分忧。

高端家具设计开始表现出对功能、空间的理解和精神层面的思考，深入洞察不同人群的需求，关注家具不同搭配后可以营造的氛围，给消费者一种精神寄托。当然，重点在于表达用户的审美和生活态度。

趋势十一：自然万物依然是灵感之源

人们向往自然，渴望住在天然绿色的环境中。这种诉求明显影响了高端家具的设计与产品生产走向，比如美式阳光海岸休闲风、英式田园风、法式田园风等，都是这种趋势的实现者。自然界里的名贵树种、各种植物的造型、自然万物的肌理与色彩等，都继续扮演家具设计的灵感之源。

趋势十二：大爱艺术笔法

高品质的艺术家具，其实就是兼具收藏价值与

使用价值的艺术品。随着物质财富的丰富与文化水平的提高，人们会对家具的艺术感提出要求。比如数百种中国文化元素的使用、欧洲文艺复兴时期的设计元素等，大有用武之地。

趋势十三：复古元素流行

复古家具的流行实际上是几种因素作用的结果，它们与一些记忆相联系，充满怀旧感。同样，复古家具也是那些喜欢绿色环保家具的人们的需要，因为它们无毒害物质的天然属性使用起来更安全，也成为很多顾客的选择。

趋势十四：中式家具回暖

随着中国GDP的高速增长、APEC峰会上中国全球影响力的展现，以及在亚太地区经济合作中所发挥的重要作用，中国正受到世界的高度关注。作为中式文化载体的现代中式家具，必然也将受到国内外消费者的追捧和喜爱。中式家具正在逐渐流行主要是由消费者观念的改变所致。一方面，与中式传统文化的逐渐认同有关，消费者开始追求传统文化等精神层面的需求。另一方面，中式家具用于收藏的观念正在改变，现代中式家具在更多的日常家居环境中使用。而消费者对于中式家具更多的需求，则又引导中式家具向简约、创意、混搭、国际化等方向发展。

趋势十五：品牌化浪潮汹涌

在高端家具市场，品牌意识的觉醒是如此的汹涌澎湃。无论是进口家具，还是本土家具，都陆续投入提升品牌影响力的工作中。另一方面，业主在选购家具时，也不再盲目，品牌因素对购买决策的影响会占据相当大的比例。

2015意大利米兰国际家具展

意大利米兰国际家具展，被誉为世界家居设计及展示的“奥斯卡”与“奥林匹克”盛会，并扩展至建筑、家纺、灯具等家居领域。2015年第54届米兰国际家具展在设计之都——意大利米兰举办，再次聚集了来自全世界的家居设计者与产业相关者。展会上的常驻知名品牌在延续了品牌个性优势的同时，进行了变化与创新。

参展品牌

ANGELO CAPPANELLINI
拥有百年家具制造历史的意大利家具品牌 ANGELO CAPPANELLINI，无疑是带有传统色彩的，在设计制作中或许会带有英国优雅绅士的老成持重和德国端庄贵族的谨慎沉稳。但卡帕奈利更多的是带有意大利的艺术风采，将南欧轻松、活泼与随意的风格发挥得淋漓尽致，在家具的设计与制作上也融入了独属意大利的贵族幽默感，令这一高档意大利奢华家具品牌成为古典森林中带有强烈时尚、灵动意味的艺术精灵。

SILIL
作为世界顶级的家具品牌，SILIK 完美象征了无与伦比的优秀品质、匠心独运的巴洛克风格、不断突破的创新元素以及卓尔不群的尊贵气质。其每一件家具的每个细节，都不容忽视地被注入了浓郁的巴洛克色彩，其高级的原材料、复杂的雕刻，彰显出皇家尊贵奢华的地位。

REXA Design
意大利品牌 REXA Design 设计的卫浴产品，干净自然的造型与现代风格的浴室相辅相成，营造出一种温暖舒适的氛围，用极致的简约之美传递出安静的力量。

CECCOTTI
CECCOTTI 用东方美学缔造出的国际著名木艺产品，其设计早已超越了东西方的界限。

ALKI
ALKI 于 1981 年诞生在法国北部的巴斯克地区的小村庄，名字 ALKI 就意味着巴斯克的意思，纯朴的现代风格缔造了另类的时尚。

BD BARCELNOA
BD BARCELNOA 延续了巴塞罗那的建筑天才高迪的智慧与创意，展现了加泰罗尼亚人独特的浪漫气节。

ALIVAR
ALIVAR 成立于 1984 年，简约时尚的设计，没有过多的装饰元素，注重实用功能与创新技术的应用。

CASSINA
CASSINA 产品蕴含了不同的语言和文化，并在风格和材料方面进行了大胆而广泛的实验。许多 CASSINA 公司的产品在问世之初由于过于前卫而受到业界的质疑。然而，随着时间的推移，这些产品最终都能为人们所接受，并成为设计的经典之作，与其他产品浑然一体。这也充分体现了 CASSINA 在作品选择上的大胆和高瞻远瞩。

用艺术眼光创造百变居庭

With Artistic Eye to Create a Changeable Patio

项目名称：中世纪之家
项目位置：美国加利福尼亚
设计公司：艺术线工作室
设计师：马克西姆
摄影师：蒂格兰

Project Name: Rodeo Drive / Mid Century Home
Location: California, USA
Design Company: Studio Artline Inc
Designer: Maxime Jacquet
Photographer: Tigran Tovmasyan

加州比佛利山庄的"中世纪"之家，以其设计跨越了时空界线。从美洲到欧洲，从美国到英国，恍然间便已至一个著名的摇滚歌星之家。卧于金属的沙发或者躺椅内，或品茶，或畅想未来都是人生一大快事。墙面定制的霓虹灯给人一种幽默般的触感。在"黑板"墙面上涂写乱想，其实是与空间的互动。白、米黄、灰褐之间间杂着现代的黑。如此快乐，令人振奋的空间，坐卧之间皆有灵感。

This little slice of mid-century heaven located in Beverly Hills, California was designed by Maxime Jacquet. Its captivating concept transports us to London England to a famous rockstars flat. We stay for tea, and jet set to the future where we lounge in metallic sofas and chairs. A custom neon sign graces the walls for a playful touch, and you can even interact in with the space and scribble on a chalkboard wall. The colors are a collection of calming yellows, whites and off whites, a bit of taupe and polished with a modern pop of black. This is a fun uplifting space, where one can sit and be inspired.

And I said
I
Love You!
SHIGERU BAN
Meier
MODERN ARCHITECTURE A-L
MODERN ARCHITECTURE M-Z
LOIS LANE
ACTION COMICS

NEW YORK PARIS
LONDON VENICE
TOKYO MOSCOW
HONG KONG SYDNEY

NEW YORK PARIS
LONDON VENICE
TOKYO MOSCOW
HONG KONG SYDNEY

DREAM
HOME Sweet HOME
Style
Dive Into sunshine
CALIFORNIA
Los Angeles
Just Love
MENU
EXPECT great THINGS
family
HAPPY
Live
PEACE
You Are My Sunshine

皇家行宫的华丽变身

Another Imperial Palace Comes Gorgeous

项目名称：阿斯科特皇家行宫
项目位置：英国阿斯科特
设计公司：伦敦 MPD
设计师：毛里齐奥
摄影师：杰克

Project Name: Ascot Lodge
Location: Ascot, UK
Design Company: MPD London
Designer: Maurizio Pellizzoni
Photographer: Jake Fitzjones

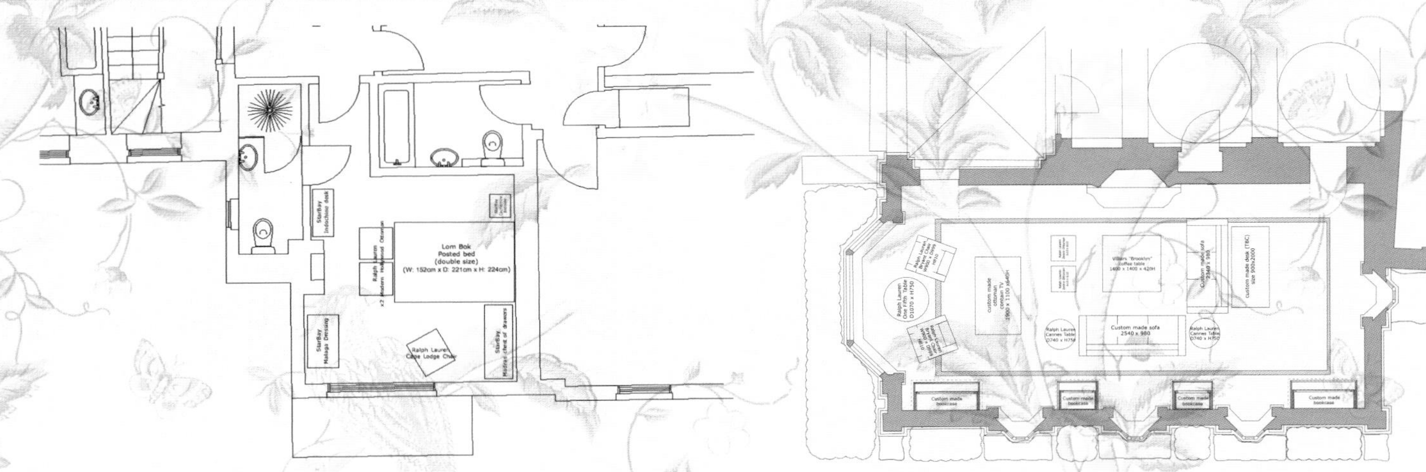

RALPH LAUREN

"阿斯科特皇家行宫"所在建筑位于英国南部伯克郡，建于 15 世纪，与英国皇室颇有渊源。国王詹姆斯一世、玛丽皇后的皇子先后把其当作自己打猎时的行宫。

该建筑室内室外几经修缮，如壁炉、地板、窗户、天花等。多次整修无不以保持建筑构造为原则。MDP 主要负责的是其中的几个关键处所，如乔治亚厅、舞厅、主卧以及两个客厅。设计嫁接的是国际著名服装品牌"拉夫·劳伦"风格的时装元素，同时糅以古董、艺品、家藏的祖传遗物。但凡历史古建筑整修时无不需要小心、谨慎，以便保持其往昔风华与印迹，本案亦然。

MDP 所经手每一个空间，皆以其中历史器物彰显其个性与特征。乔治亚厅如今华丽转身成了一个私人画廊。壁板的恢复可谓是精心竭虑。其上的电源轨焕发出了新的活力。空间铺陈的艺品无不是设计师约翰·琼斯亲力亲为。所有元素耗时耗力。

一应铺陈是"拉夫·劳伦"品牌的大荟萃。其中有些家具、古董是出自 MPD，有些为家传珍宝。柔软的织物、窗帘源于 De Le Cuona。温馨的家居空间，富有历史底蕴的环境就这样深深地烙印上了"个人品位"。

This Grade II listed building has a fascinating history; it was built in the 15th century and was used by King James 1, son of Mary Queen of Scotts as a hunting lodge in the Berkshire countryside. Modern day alterations to the interior and exterior were limited due to the fact it is a listed building and all the original fabric had to be maintained. This included fireplaces, floors, windows, doors, ceilings, cornicing, coving and all other architectural detailing. MPD worked on the refurbishment of the interior of some of the key rooms in the house: the Georgian Room, Ballroom, Master Bedroom and two guest bedrooms. The brief was to draw on elements of Ralph Lauren styling but also to incorporate antiques, artworks and personal heirlooms belonging

to the family. This unique property required sensitive, thoughtful treatment bringing it up to date whilst retaining character and charm.
Each room has a signature style inspired by a piece of artwork or connected to the history of the room itself, as in the case of the Georgian Room which was transformed into a private gallery with the use of a power rail attached above the original paneling which was carefully restored. All of the artwork was framed by master framer John Jones and was a lengthy and meticulously planned element of the project.
The furniture is a mix of Ralph Lauren pieces, bespoke furniture and antiques, some of which were sourced by MPD and others belonging to the family, which they wished to incorporate into the design. Soft furnishing fabrics and curtains were sourced from De Le Cuona. The intention was to create a warm family home which would grow and develop with the owners bearing the stamp of their individual taste and family history.

EARTH

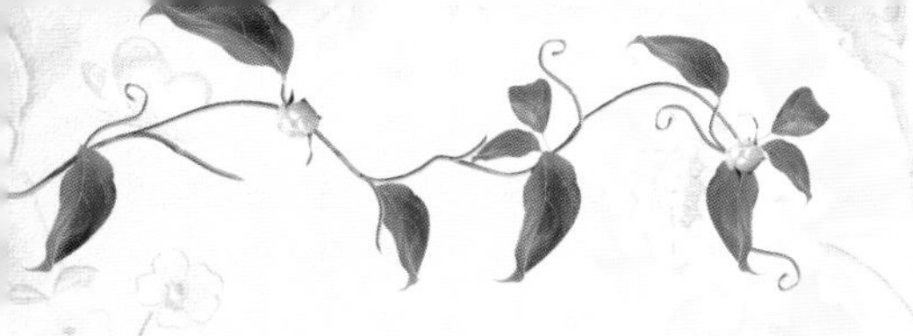

弘阔大海是豪宅窗前最美的装饰

The Best Window Decoration: Sea View

项目名称：圣特仑塔
设计公司：安多工作室
建筑师：阿齐扎
设计师：爱泼斯坦

Project Name: Trumpeldor St. Tower Apartment
Design Company: Ando-Studio
Architect: Aziza
Designer: Epstein Anna

本案公寓位于以色列港口城市特拉维夫，地处地中海岸边，城市天际景观尽收于眼。无论是细部设计，还是家具位置、用材饰面、木材、地板风格等全部经过 3D 模拟测试。其中家具、艺品、配件等由业主亲自精心挑选。
如何使空间最大化地享受到大海的美景是设计的重中之重。混凝土、木材、钢材一并使用。黑白色系用于室内，但却连接着内外之间。奢华的风格，使空间保持着高水准，利于空间销售。

The apartments are located in Tel Aviv, Israel. The location of the tower is in the city sea promenade, first line to the Mediterranean Sea shore. The design process included a lot of 3d tests of details, positions of furniture, finish materials such as walls colors, woods, flooring styles etc. It also included a lot of work with the client for choosing the right furniture, art and accessories.
The design was trying to keep the focus on the view to the sea. The materials used were concrete, wood, steel and white. The colors of the tower are black and white and we tried to use them inside the apartments to keep the connection between in and out. As all our projects we use the latest and high-end furniture and accessories. It was very important to keep a high level for those apartment since the apartments planed as a very luxurious units for sale.

以不同坐姿，体会空间的变化乐趣

Sitting in Different Gestures to Perceive Fun

项目名称：松林公寓
项目位置：巴西圣保罗
设计公司：AMC 建筑设计
设计师：安东尼奥、马里奥
摄影师：马克
面积：493 m^2

Project Name: Residence Pinheiros
Location: Sao Paulo, Brazil
Design Company: AMC – Architecture and Interiors
Designer: Antonio Ferreira Junior, Mario Celso Bernardes
Photographer: Marco Antonio
Area: 493 m^2

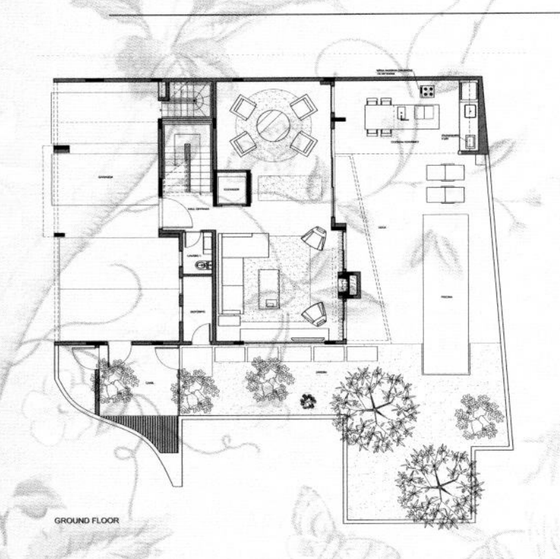

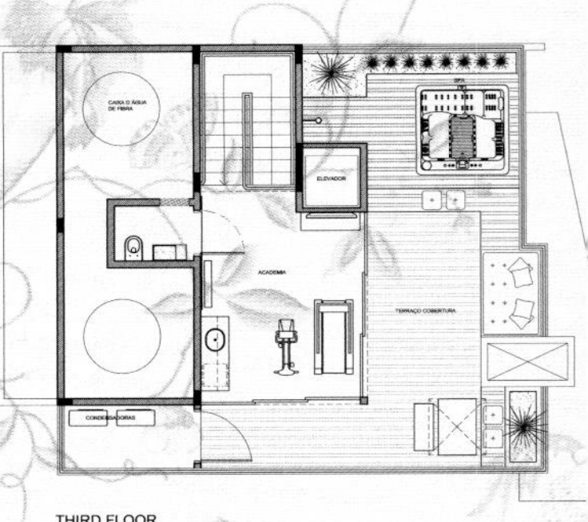

“松林公寓”是8栋别墅中的一栋。除了与其他空间共有的元素外，空间不失优雅与舒适。虽然贵为别墅，但公寓内部应有的陈设应有尽有。现代的线条，色彩斑斓的世界。黑色的地板砖分布于所有空间，白色漆面，灰色、银色的墙面，给人一种安静的感觉。各色家具的铺陈彰显出现代格调。

The residence located in a condominium of eight houses, the apartments have common elements, without losing sight of the charm and comfort of a home. Despite being a house, she has all the characteristics of an apartment because almost everything happens inside. Following a very modern line, the décor uses and abuses of color spaces. Sobriety was due to the black porcelain floors, distributed in all environments, and the application of white paint, gray and silver walls. The colors were distributed in contemporary furniture.

AMBIENTES
RUNNERS
Guia

法式折中主义的时尚新装

Fashion of French Eclecticism

项目名称：伏尔泰塔
项目位置：美国加利福尼亚
设计公司：艺术线工作室
设计师：马克西姆
摄影师：蒂格兰

Project Name: The Voltaire Tower
Location: California, USA
Design Company: Studio Artline Inc
Designer: Maxime Jacquet
Photographer: Tigran Tovmasyan

本案所在建筑原建于20世纪30年代，法式折中主义风格，建筑名为"伏尔泰塔"。7层的大楼，共40户。始建时，便吸引了一大批名流入住，如好莱坞红星安·萨森等。玛丽莲·梦露1954年离异后曾在此短暂居住，盛传其所居住的房间便是本案。80年代，空间一度改建成奢华宾馆，几年后再经流转，成了迎合上流社会消费的所在。好莱坞的明星大腕更是把此处称为"宫殿之家"。

The Apartment is located in the Granville Towers, which was originally an apartment building named The Voltaire, it was built in 1930 in the French Revival style by architect Leland Bryant. The 7-storey, 40-unit property was a celebrity magnet from the very beginning and such stars as Ann Sothern, Jack Lord, Arthur Treacher, Janet Gaynor, and Rock Hudson called the place home. Marilyn Monroe even stayed there for a brief while after her divorce from Joe DiMaggio in 1954. The rumor of the building is that she lived in the unit 61 that is the unit I designed and lived in. In the 1980s, the property was transformed into a luxury hotel at which point it was renamed The Granville. A few years later it was transformed yet again, this time into an upscale condominium building, and Hollywood luminaries once again began calling the place home.

PORSCHE

HELMUT NEWTON
HELMUT NEWTON

ARAKI

HELMUT NEWTON
HELMUT NEWTON

YVES

CHANEL
TERRY RICHARDSON
TERRY O'NEILL

TOM FORD
HADID

亲密小空间，层层有惊喜

Each Floor Intimate but Small for Surprises

项目名称：玛丽安娜别墅
设计公司：古托工作室
设计师：古托

Project Name: Vila Mariana
Design Company: Guto Requena Studio
Designer: Guto Requena

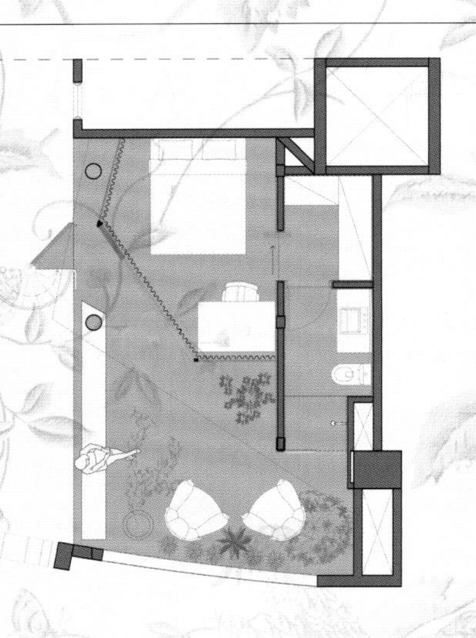

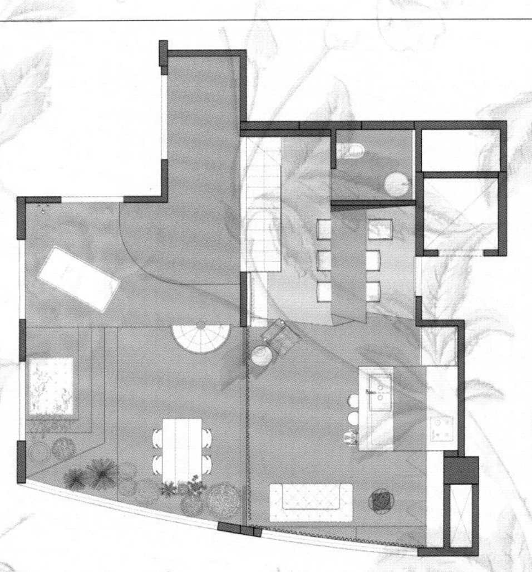

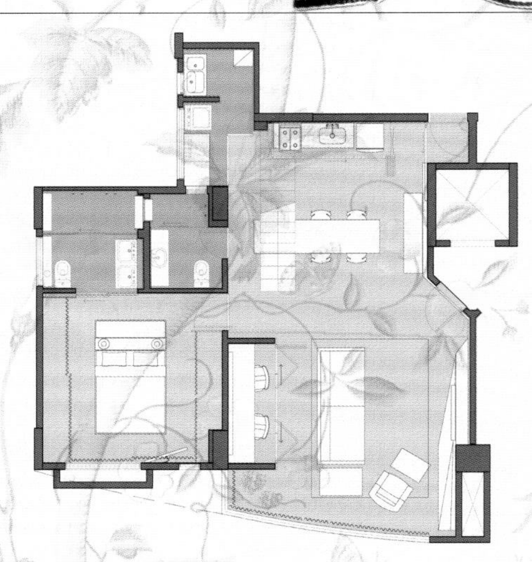

“玛丽安娜别墅”位于巴西圣保罗。三层的空间借助于设计妙手终成了一个终极的现代家居空间。如混凝土、木材等各种现代的用材与饱和的色彩共同铸就了空间的华丽。设计后的空间，玄关位于二楼。曾经的厨房、客厅、两个卧室如今成了美食厨房、别致的生活区。新生的令人愉快空间就这样自然而然地成了整个空间的一份子。

房间般大小的阳台悬挂着两个太阳椅。太阳椅的位置正好靠着泳池的边缘。泳池边壁涂以黑色，隐私感极强。另有边壁正好依托着建筑的立面。而最后一个边壁与旋转楼梯下相连。沿楼梯攀援而上，抬眼便是茂盛的热带植物。两把标志性的椅子，着以橘色的涂装，分外耀眼。椅子巴洛克的细节对比着建筑现代清爽的线条。明亮非凡的色调正好辉映着原本极其传统的建筑轮廓。混凝土的朴素，自然的绿色恰成空间的背景。与露台相邻的是客房。客房虽开有落地窗，观景角度好，但隐私感强。

小小的洗衣房位于后部，正好位于客厅的后边，厨房的旁边。大大的蓝色锣鼓在此演化成盥洗池的底座。绘有热带花鸟的墙纸悄然而立，但却对“锣鼓”起着极好的衬托作用。

另有小厨房，气氛亲密，是一个不可多得的备餐场所。火炉与冰箱之间

有2.3米的距离，足以安装下定制的混凝土岛柜。
办公室同样位于客厅的旁边，但却隐于壁板之后。与办公室相邻的是主卧。卧床的床头墙高低参差不齐。矮墙的出现其实是对空间的界定。卧室有些小巧，以炭黑色帷幔作为装饰。淋浴室装饰着紫色的瓷砖。白色的混凝土、大大的风景画彰显着设计的别样特色。全高的墙面背后是化妆间。化妆间色度幽暗，富有张力，与三层的另外两个卫生间形成鲜明的对比。

Architect Guto Requena was challenged to convert a triplex located in Sao Paulo, Brazil, into an ultra-modern single family home that combined the use of saturated color with contemporary materials such as concrete and wood. The new main entry is locate on the second floor, and here - what was originally a kitchen, living room and two bedrooms - is now a gourmet kitchen, and a unique dining space for entertaining that melds into and becomes part of the building structure.
On the platform there is just enough room for two green sling back sun

chairs before the edge of the pool begins. The pool is surrounded on two sides by black privacy walls, the third side is the concrete facade of the building and on the fourth side is the beginning of the spiral staircase. The spiral staircase leads up to a private terrace of lush plants and two statement chairs covered in a bold choice of tangerine. The Baroque detailing of the chairs is a well-planned statement against the clean modern lines of the building. The combination of a bright and unusual color on a more traditional silhouette is completely on point with current trends as is the abundance of natural concrete and natural greenery. Adjoining the terrace is the guest room. The room offers a space of complete privacy from the rest of the home as well as spectacular views easily seen through the abundance of floor to ceiling windows.

Back down on the second level there is a small washroom beside the kitchen and behind the dining area. Here, a large blue drum converted into a pedestal sink is juxtaposed against fun wallpapering of blue tropical birds and jet black detailing.

A second, smaller kitchen offers a place to prepare meals for more intimate settings. Here there is only 2.3 meters of counter space between the stove and fridge, but there is plenty of prep space available on the custom multi purpose concrete island.

Just like the kitchen, the office space behind the living area can be hidden behind a series of wall panels. Behind the hidden office is the Master Bedroom. The headboard wall of the Master Bedroom is actually part pony wall, part full height wall, with the pony wall creating a division to the ensuite. The bedroom is small in size and wrapped in charcoal drapery. Color is introduced by the purple tiles used in the walk in shower. The tiles are highlighted even more with the use of white grout and the large framed photo on the wall of a scenic day is another carefully planned use of opposing design features. The full height headboard wall in the Master Bedroom is the back of a small powder room. This powder room is dark and dramatic, a complete contrast to the mood of the other two washrooms within the 3 levels of this home.

CHANEL
BRAZILIANartBOOK

RAYMOND CHANDLER
RAYMOND CHANDLER
JANELA PARA A MORTE

热情待客，舒适居家

Home: Welcoming and Comfort

项目名称：西单地大公馆
设计师：格伦
摄影：GD 摄影提供

Project Name: West End Avenue Residence
Designer: Glenn Gissler
Photography: Gross & Daley Photography

业主近20年居于新泽西，希望能对该处原建于战前的房子进行修缮，在快乐、健康、富有魅力的环境下，一家人其乐融融地开始生活的新篇章。
“型状”的生活空间里尽优雅与舒适，但却无任何虚饰。无论是独居于空间，还是与家人共同生活，得到的总是舒适。偶尔有客来访，餐饮、小住，宾主相尽欢。逢有大型的家庭聚会，也是低调地为客人提供着放松与享受。
他处生活中的家具在本处空间里出现，是内敛，也是业主生活价值观念与经历的反映。无论你是客人，还是主人，令人放松的布局、暖色的调色板给人的感觉除了放松，还是放松。所有家具铺陈，不同文化、不同时期、异样风格考验着那些极具甄别目光，受过良好教育的人士。精良的艺术陈设于半空阔、半充实的空间中书写着本案的美学格调，真的很令人快乐。

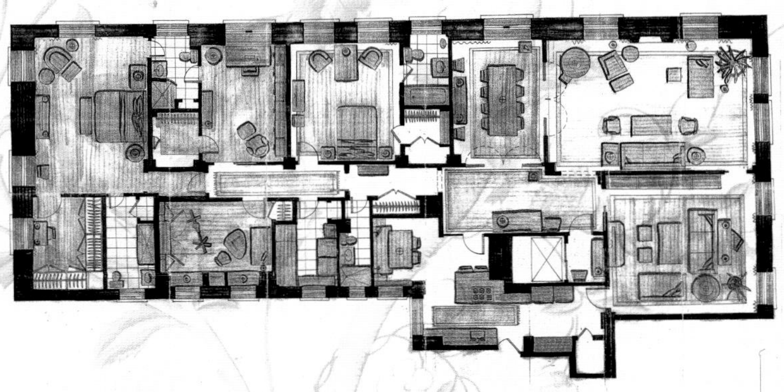

A graciously scaled pre-war apartment was renovated, and decorated for happy, healthy, and attractive empty nesters returning to New York City after two decades in a New Jersey suburb ready to create the next chapter of their lives together. A lifestyle program was developed to use all the many spaces of the apartment with grace and ease, and a lack of pretense. Comfortable spaces were created for each of the occupants in numerous spaces around the apartment for time alone and time together. Visiting family is easily accommodated for casual meals, and short or long stays. The apartment makes entertaining many guests for more formal dinners or larger crowds easy and gracious while maintaining understatement, still providing opportunities for delight.

de Kooning
BERENICE ABBOTT CHANGING NEW YORK

While this new home integrated few furnishings from their 'previous life' an understated yet stimulating environment was created that successfully reflected their values, interests, and travels. Guests and owners alike feel comfortable in the new home due to the easy layout, and warm palette. More educated eyes take delight in the diverse array of cultures, time periods, styles represented in the furnishings and objects throughout the apartment. Acquiring and developing a fine art collection was a key element and enjoyable aspect in the making of a new home that is aesthetically both half full, and half empty, but experientially a total joy.

春色

Colors of Spring

设计师：尤里
面积：100 m²

Designer: Yuri Zimenko
Area: 100 m²

"春色"的风格是自然成风：美式新古典。折中的开始糅合进些许自由。
传统的苏格兰格子背景下，刮起了美式新古典的风。无论是软装的椅子、沙发还是定制的羊毛地毯，都彰显着一种酽酽的美国风。家具是建于1885年的意大利Nino Galimberti品牌。该品牌多以木质古董家具而出名。厨房、浴室大理石板与卧室里的黑色胡桃木形成鲜明的对比。墙面灰蓝色调。米黄、黄色轻轻地柔化着稍冷的色调。如客厅里的珍珠色，让视野有了一种生动的变化。

The current interior solution the designer calls the American neoclassicism. In his opinion, it carries an eclectic beginning and at the same time has a certain degree of freedom.
The main decorative element became a figure in traditional Scottish tartan check. It can be found here everywhere from the upholstery of chairs and sofas to custom made luxury wool carpet in the living room. Amazingly accurately the character of the room furniture underlined one of the oldest Italian factories Nino Galimberti (1885), famous for its unique work with wood. Marble is selected for the floor in kitchen and bathroom, and for the bedroom - dark walnut. The walls are painted in gray-blue tones. Shades of beige and yellow slightly soften the cool palette. For example, in the living room, to coating added: pearl, due to which an interesting visual effect appeared.

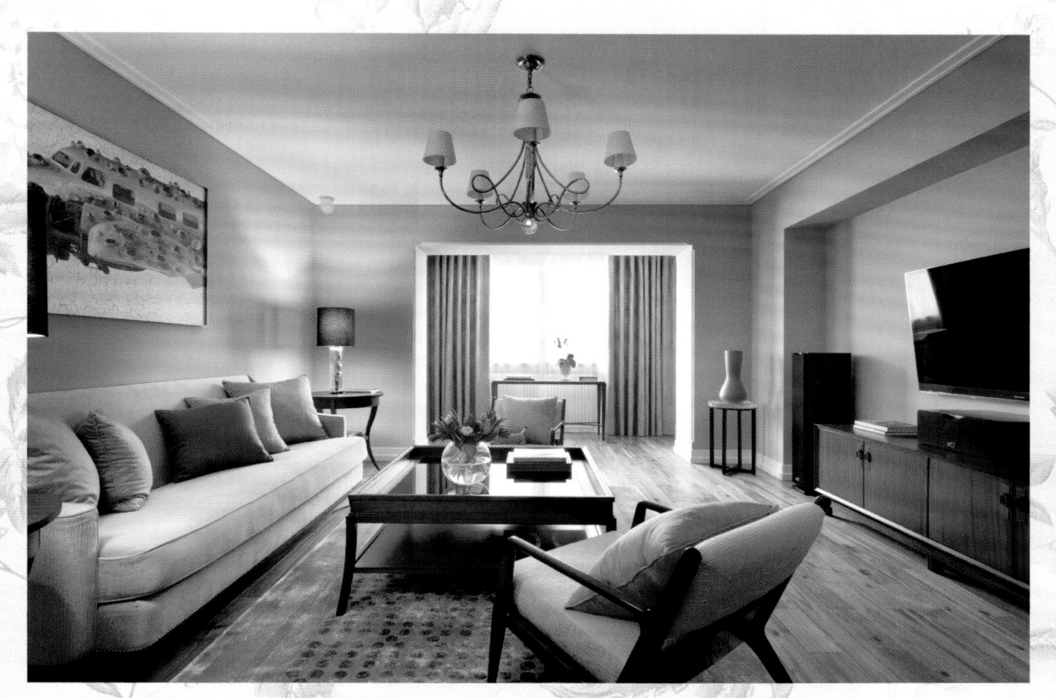

INTERIORS

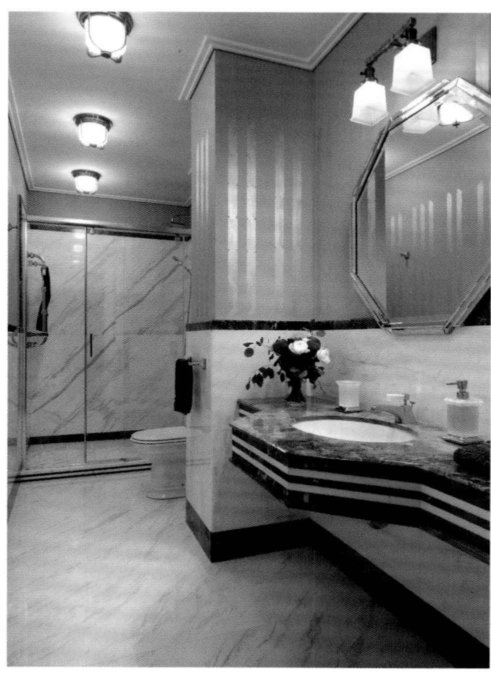

尊重历史，崇尚自由

History Respected, Freedom Adored

项目名称：里瓦罗洛别墅
项目位置：意大利
设计公司：彼得罗建筑事务所

Project Name: Villa Rivarolo Renovation
Location: Italy
Design Company: Pietro Carlo Pellegrini Architetto

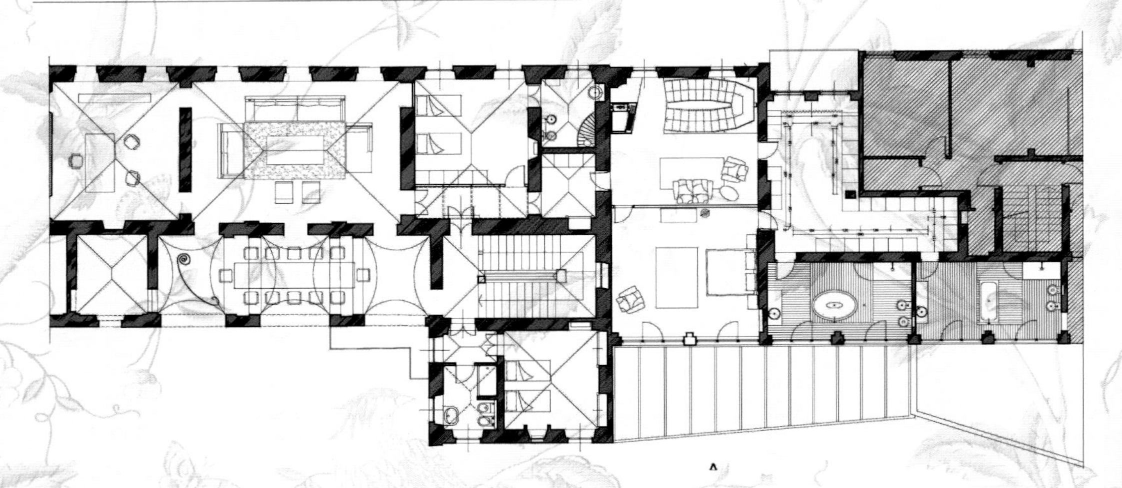

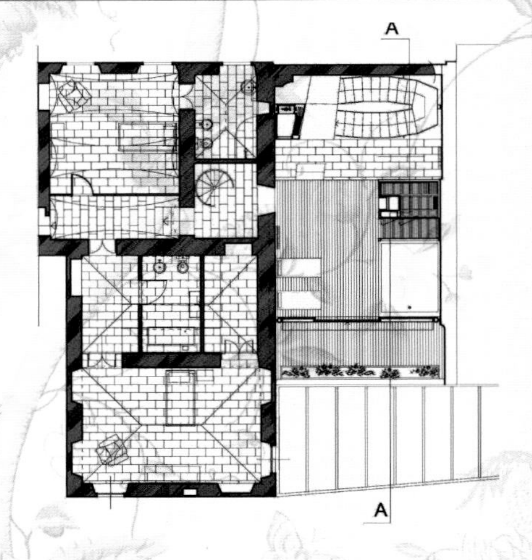

"里瓦罗洛别墅"的升级，首先从对其历史进行精准的分析开始。在本案看来，本着对历史的尊重，解读建筑本身及其相关的历史特征是设计的起点。地板的修复及强化，同时有助于灯光的使用及对透明结构的利用。

为了使量体的面积得到更好利用，空间的水平、垂直结构在某些方面进行了重置。屋顶、地板、室内装饰等修旧如旧。这不仅是从历史的角度，同样从功用的角度出发。同时，空间添加了悬挂钢构楼梯，着以白色，连接着新增添的建筑与旧有的建筑。楼梯台步由灰色石材铸就。除了新加楼梯，空间还加了一个小巧的升降梯。升降梯的四壁随机地点缀着圆形的窗户，强化着空间的垂直视觉感。

因为灯光、玻璃的共同渲染，整个空间给人一种通透、明亮的感觉，彰显着量体外观的纯洁与设计细节的考究。

空间用材精心挑选，尽显室内终极形象。古董映衬着现代的铺陈。现代的铺陈又与经典悄悄地进行着互动。家具摆设自然而然地融于新旧合二为一的空间中。

The renovation follows an accurate historical analysis, which studies the evolution of the building. The reading of the building own architectural and historical features is the starting point for the architectural action, which respects the original trademarks (i.e. floor restoration and strengthening) but also provides for the use of new light and transparent structures.

The project includes the reorganization of the house from a functional point of view; in order to better exploit the available volume, the horizontal and vertical structures has been partly reset. The roof and floors has been restored, as well as the internal coverings and decorations. A new area of the building has been totally renovated and connected to the existing part of the building through a new white painted suspended steel stairs, with grey stone steps (Matraia stone); a new small elevator has been added, and its surface, which underlines the verticality of the space, is spotted with random circle windows.

The main feature of the project is the use of light materials and glass walls, which, together with light colors, give a sense of brightness to the entire house, underlining the

purity of the shapes and the care of details.
The use of materials is carefully studied and it's fundamental for the final image of the interior. The result is a beautiful combination of antiquity and modernity, where modern materials interact with classical decorations. The furniture, as well, has been carefully chosen and becomes part of this union between new and old.

灯具流行趋势

Trend of Lighting Popularity

现代家具设计与现代灯饰设计正逐步融为一体，这是灯饰设计面临的新课题。现代家具与现代灯饰的整合化从20世纪90年代以来已经成为现代家具的大趋势。全球著名的意大利米兰家具国际博览会、德国科隆国家家具博览会、美国高点国际家具博览会都是把家具与灯饰作为一个整体系列来设计、陈列、展览与销售的。越来越多的设计师对家具与灯具的整体配套设计表现出极大的兴趣。众多元素的组合作用让现代家具的形态大放异彩。家具要讲究造型、结构、材质肌理等总体形态效应；灯具应讲究光、造型、色质、结构等总体形态效应，两者都是构成建筑环境空间效果的基础，它们互为衬托，交相辉映。

各式各样的灯具除要满足照明的需求外，亦要符合装饰和美化居室的效果。现在人们在装修家居时更加注重细节，哪怕是浴室镜子上方的一盏小灯，也追求时尚和别致。当家具与灯饰越来越潮流化的时候，灯饰作为与家具、装饰互相搭配的一个重要元素也随之发生了变化。头顶上的吊灯、茶几和书桌上的台灯已经悄然“变脸”，颜色变得更加多彩、清新，材质愈加清透，造型上也显示出了天马行空般的创意，甚至在一些家具城里，抢眼的灯饰俨然成了主角，吸引着人们的注意力。

随着生活水平的逐步提高，人们越来越向往较高的生活品质，因此对灯饰也不断提出了新要求。纵观近年来的灯饰发展，主要表现出以下趋势。

流行趋势一：结构简单、做工精细、色彩明快

现在，家居装饰用灯设计的潮流越来越趋于结构简单、做工精细、色彩明快，正像我们在北京看到的越来越多的意大利家具，他们的设计风格十分现代。而体现着当今世界高档时尚风格的家居用灯，也随着现代风格的家具走入中国家庭，也正是具有此种简洁化设计风格的灯具，才更加注重设计简洁，注重合理、充分利用光源的照明效率与家具交相辉映。而并不像传统概念中那种结构复杂、色彩斑斓、体形笨重、设计忽略光效充分利用的灯具。这是一种设计观念的革命。同时，这种高档时尚灯具的设计也符合节能要求。

流行趋势二：美观、实用、个性化

理想家居装饰灯的概念将是美观、实用、个性化。强调自我、追求个性将成为许多顾客的首要选择。市场上家居与灯饰的日益丰富为顾客提供了更多的选择，而顾客爱好取向的个性化又对家具与灯饰的发展提出了更新更高的要求，促使家具与灯饰不断结合。

流行趋势三：文化含量高、以人为本、节能环保

随着新一代高文化素养的消费群体的增大，家居装饰的文化含量将大大提高，在家具配置方面，能体现文化底蕴的灯饰将大受欢迎。当人们在追求家居生活舒适性的时候，人性化设计是满足这一需求的解决之道，无谓与花哨的装饰将遭淘汰。无论是装饰或是家具，都以“以人为本”的概念作为设计的指导思想。在当代社会，随着人们环保意识的增强，环保设计和以环保材料制作的家具会越来越多。在今后的家居装饰中，具有环保功能的灯饰会得到越来越多的运用。智能化也是今后家居装饰的重点发展方向。家里的家具、门窗、照明器具、电器、厨房卫生间用具等，都将根据不同使用者在不同时间的不同需求作相应的、智能化的配置，满足现代人的需求。

流行趋势四：多光源效果

目前，国内的室内照明设计已由过去仅注重单光源过渡到追求多光源的效果。这样的变化表明，设计师已经意识到良好而健康的灯光设计对人们生活的影响。在单光源时代，客厅和卧室往往由一盏灯统领全局。而现在，多光源设计已经照顾到每一个使用者和每一种生活情境对灯光的需求。主光源提供的照明使室内都有均匀的照度；而展示灯、台灯等提供的重点照明或局部照明，则丰富了空间照明的层次。多光源的配合使得空间照明无论是浓墨重彩还是轻描淡写，都能形成曼妙的空间氛围。

流行趋势五：灯饰艺术感带装饰性产品走俏

从 2015 年米兰国际灯饰展可以看出，其设计风格不仅重视古典原貌，也强调艺术自然化。有古色古香的灯饰，其灵感来自 18 世纪欧洲古建筑的风貌，其外表精美，做工细腻，带有浓郁的欧洲宫廷气息。而今年带有现代风格设计的灯具，更重用材，采用现代高科技的新材料、新工艺和新技术，不仅有照明功能，还带有装饰性，可和墙体以及其他物体结合，起到家居装饰的作用。

流行趋势六：科技主义

“只要好的，不怕贵的。”灯具市场刮起一阵“科技风”，新材料、新工艺、新技术不断涌现，让灯具从一个照明产品变成彰显主人身份的高科技产品，即使价格贵一些、造型差一些，也因为它高贵的材质或玄妙的技术而被人们有意无意地忽略。

纳米、光纤、LED、手触摸、远程遥控……这些看起来是与 IT 或高科技领域相关的技术词汇，本与灯具行业似乎风马牛不相及，但在近年里，却成为盛行于灯具市场的流行词。随着技术的进步，如今的灯具可以采用光纤材料制造，营造特别的氛围；可以用 LED 发光源代替传统发光源，使光线更柔美，产品寿命更长久；可以利用高科技手段实行远程遥控或只需用手轻轻触摸即可实现灯具的开关与亮度控制，轻松提高人们家居生活的舒适度和享受度。

或许这种灯具价格更高一些，但对于追求生活品质的人们来说，却是越高档，越新鲜、越具科技含量，越是愿意追逐。“科技主义”正在灯具设计中悄然流行，置身家中的你如果想犯懒，不妨好好研读包装盒里的说明书，操控一下家里的灯具，享受“科技主义”流行趋势带来的便捷与神奇。

灯具名目繁多，包括水晶灯、手工玻璃灯、LED 艺术节能灯等，按用途来分类，包括吊灯、吸顶灯、落地灯、台灯、庭院灯等，材料有木材、金属、塑料、纸张、新型复合材料、化纤、陶瓷、硅胶等。

水晶灯

2015 年，水晶灯将继续向时尚现代方向倾斜，款式和风格都倾向多样化，装饰效果越突出越有市场。2015 年，水晶灯材质组合更加丰富，金箔、铜材、琉璃、玉石都成为水晶灯的组合元素，不仅雍容华贵，而且时尚大气。传统黄水晶份额继续缩小，但与佛教情境和华丽美学相结合的经典款式依然畅销。

LED 灯

近年，LED 作为高效、低耗的新兴光源发展十分迅猛，诸多业内人士纷纷进行 LED 取代高压钠灯方面的研究和探索。随着半导体照明产品性能的持续提升和制造成本的不断降低，同时也随着相关应用环境，如标准、产品认证和检测、工程应用设计和示范、相关政策等产业环境的完善，LED 照明应用将在 2011—2015 年间保持较高的增长速度。按照灯具数据计算，预计 2015 年其在照明市场的份额将提升到 30% 左右，成为最具竞争力的主流照明光源之一。

此外，中国互联网的崛起，将带动 LED 产业的崛起，这是半导体产业中，中国与国际先进水平最为接近的领域。微电子和光电子携手并进的时代来临，而 LED 照明与信息技术深度融合（Li-Fi），也将成为大数据时代的重要元素。

彩绘玻璃灯

彩绘玻璃灯具色彩丰富，造型生动，表面细腻光滑，极具装饰性，堪称精湛工艺与高雅情趣的完融合。

彩绘玻璃灯具从装饰手法上可分为彩绘、手绘和马赛克粘贴三种。彩绘手法是将玻璃切割后包上铜箔，再经焊接而成。手绘玻璃则采用磨砂效果的雾面玻璃和特定进口颜料，先在玻璃内画图，再经高温烘烤，反复三次，最后贴上固定的保护膜而成。马赛克粘贴指在灯罩表面罩以彩绘的碎玻璃，立体感强烈。但由于碎玻璃每一篇均需手工镶嵌，故而相当耗工费时。

彩绘玻璃灯具颜色永不改变，晶莹灿烂。远看或开灯后更能体会到它的高贵和华丽——潜藏在暗处，只需一点光，在折射和反射的往来中，熠熠生辉，绮丽动人。

光是看得见却又摸不着的神奇存在，从光中可以感受到时光、距离、情感。作为米兰家具展中不可错过的展览，2015 年米兰国际灯饰展（Euroluce International Lighting Exhibition）汇聚当下潮流趋势，亮相的材质及色彩为灯饰用品带来与众不同的触感和深度感视觉盛宴。在这里，来自全球和意大利本土的顶级灯饰品牌参展，让灯具的内涵和外延都有了深层的拓展。

米兰有了光，世界就有了光。以下是我们从2015年米兰国际灯饰展中甄选出的无论是设计、材料、造型、还是光线都美得让你惊艳的灯饰设计作品：

1. 雅致瓷偶（Lladro）全新“法兰西小姐”灯具系列：雅致瓷偶的灵感来源与如何将其经典的女性雕像塑造成兼具美观的功能性作品。2015年米兰国际灯饰展推出的全新灯具系列包括四件独立吊灯和三件枝形吊灯，均配以9只、18只及25只灯泡与4种模式相结合的灯光。柔和的灯光透过“法兰西小姐”们的陶瓷裙摆如氤氲般梦幻。与此同时，陶瓷表面雕刻精美的装饰图案也得到了充分地展示，二者相得益彰，完美诠释了女性永恒的魅力。

2. Notch：由Michael Anastassiades设计的Notch是灯光和珠宝的结合，重视并增强美感。它定义了“挂件”的含义，同时讲述着两个世界：一件戴在颈上的项链和用来挂在天花板上的光。

3. Ether：由Philippe Starck设计，这是一个新的概念：一个空灵、纯净、几乎是无形的灯。她赤裸裸的，有着透明的基座，可以根据个人的口味和风格搭配铬或铜的灯帽，而灯罩部分则又非常丰富的选择，比如像磨砂玻璃一样的琥珀色注塑塑料的褶皱织物。为她更换灯罩，就好像为一位美丽的女士更换衣服，性感而随意地表达万种风情。

4. Hollyg：Hollgy由Giorgio Biscaro设计，是LED灯与玻璃吹制艺术的合体。尽管是传统的材料和工艺，Hollyg却包含一个重要的技术创新：LED光静静地打在整个吹制玻璃上，而可移动的盘灯罩又将其反射回弯曲的玻璃表面，使人不知道光从哪里来，好像是无中生有的神秘。

5. Slamp：服装设计师的职责是将平面布料创造出穿在人身上的三维效果，现在，同样的理念也被运用到Slamp灯具的设计上。只不过，当设计的目的由装扮身体变为装点空间看，那么，能达到三位效果的材料便只有一种——光。Slamp通过材料的剪、折、穿孔、断层架构，使得不可捉摸的光线拥有了迷人的形状。

6. Kushi lamps：名为Kushi lamps的灯具，由设计师Alberto Saggia和Valerio Sommella共同设计，可做为台灯和吊灯使用，由两层人工吹制玻璃制成。无论是它的外形、光滑的曲线还是特制的金属柄，都让人联想到苹果。

7. Blum Floor Lamp：设计师Arturo Alvarez设计的Blum Floor Lamp，其“座右铭”是：并非所有事物都是表面的样子。令人惊奇的是，这款灯具由钢丝制成，看起来坚硬冰冷，但它的光线十分柔和、温暖。

8. Icarus Wall Lamp：同样是设计师Arturo Alvarez的作品。这款灯具看起来像一个太阳圆盘，非常引人注目。它的光线十分强烈、均匀，全方位地照亮空间。

9. Stochastic：由丹尼尔Rybakken设计的经典吊灯，装饰美观，散发着精致的气息，着实令人回味。微妙的硼硅玻璃球体，从日常视图隐藏光源，它发出的光变为该组合物的主要部分。

10. Edison's Nightmare（直译为“爱迪生的噩梦”）：这款产品的设计师是Harry Thaler，他的作品是一款被钉在墙上的有着白炽灯外形的灯泡。这款艺术品象征着照明设计和技术的变迁。

11. Stones：为什么这块巨大而沉重的石头竟漂浮在空中？这是设计师Andrea Bastianello利用玻璃纤维制作的人造宝石灯罩，给人一种石头的错觉。

12. Supernova：意大利建筑师Ferruccio Laviani的作品。这款吊灯提供360°的背景照明，采用了14块圆形铝盘和钢盘，外形看起来就像带有百叶窗的球体。

13. Life-Size灯具系列：Life-Size雕刻灯具系列来自于LZF Lamps公司创建人Marivi Calvo和Sandro Tothill的合作，在此，艺术家和工匠展开跨领域的协作与突破创新，向对独特物件感兴趣的人们呈现出集艺术和设计于一身的新创意。锦鲤灯具是Life-Size灯具系列中的第一件，是独特雕刻创意中的一个创新系列，它与插画家Isidroferrer的“有趣的农场动物”、Burkhard的“蒲公英”等其他物件一起出展。

14. Kurage：日本设计工作室Nendo和意大利设计师Luca Nichetto联合为威尼斯灯具品牌Foscarini推出的这款纸质灯具拥有一个奶油冰棍般清爽的外形，日本桧木打造的支架如同冰棍一般，和纸灯罩结合日本传统的纸张印染技术和现代的三维工艺，极其轻巧地构成了半圆形的穹顶。当灯光亮起，让人想起水中的水母。

15. Orient：这款吊灯早在1963年就由设计师Jo Hammerborg设计完成，如今其家族成员推出了新版本，采用了闪亮的铜外罩，紫檀木做顶。

16. Acqua：这款灯具是设计师Luta Bettonica的作品。它的外形独特，灯光发射面是一个弯曲的类似水滴的凹面。每只灯具还可以一个连一个的组合到一起，形成不同的外形。

17. Formala：同样是设计师Luta Bettonica的作品，由维特拉设计博物馆（Vitra Design Museum）推荐参加灯具展。Formala由灵活的金属条组成，可以固定到不同的位置，变换不同的形状。

18. Golden Ring：Golden Ring是一款伟大的设计，它的外形独具匠心，并且扩大了照射范围。它是一款铝制灯具，由四个直径分别为770、1200、1800和3000厘米的圆环组成。既可单环使用，也可由不同大小的圆环组合成不同形状。

19. Lum：设计师Estefania Johnson的作品。灯具材料为天然氧化铜，它的外观看起来就像一面变形的墙。

20. MELT Lamp：如今最炙手可热的英国鬼才设计师Tom Dixon此次家具设计周上推出了他和瑞士团队Front合作的Melt口吹玻璃吊灯系列，这些吊灯看上去就像是我们小时候玩的口吹泡泡一样，形状随意自然，发出的光有的柔和温馨，有的变幻鬼魅，朋克气氛十足。Tom Dixon此次把展览地点选在了米兰一个荒废许久的歌剧院Casa dell' Opera Nazionale Balilla，配合其自身和展览的气质。

色彩盛宴

Feast with Colors

项目名称：布鲁克林联排别墅
项目位置：美国纽约

Project Name: Brooklyn Heights Townhouse
Location: New York, USA

5 层的联排别墅，年代久远，建于 1836 年。面积达 485 平方米，在当地极具里程碑意义。初建时的立面，19 世纪的室内设计几经风雨，但却保存完好。借助于现代的建筑手法，修复后的空间再展生命活力。厨房、客厅之间的界定得以拆除，空间面积不仅得到了放大，更让家居有了现代的社会功能。曾经的两个卧室如今合为主卧，空间的紧凑感得到了加强。厨房使用的瓷砖，穿室外阳台，直入后花园。后花园因此成了室内的绿景。空间内外的沟通与交流也得到了进一步增强。整个空间如同一个变幻多彩的调色板。折中的各种家具，为传统、现代的空间增加了层次感。

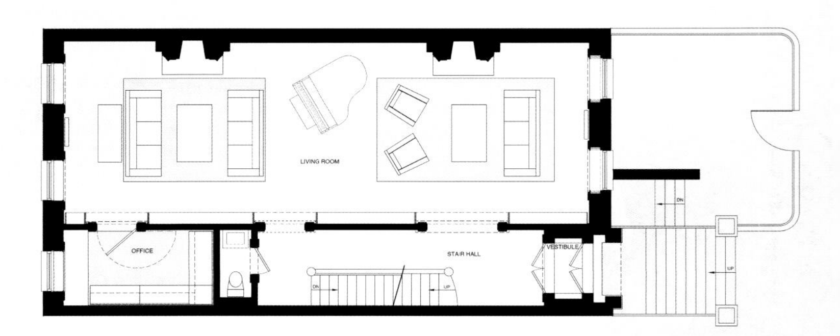

This 486 square meters five-storey landmarked townhouse was originally constructed in 1836. The house's traditional facade and intact 19th century interior detailing have a visual counterpoint in the new modern architectural and technological interventions of this renovation.

Interior partitions were removed from the living room and the kitchen, enlarging both rooms' and allowing modern social and functional uses to be integrated into the home. Two bedrooms were combined into a full master suite, creating a coherent unified space. The rear garden was designed as an extension of the interior's aesthetic with new porcelain floor tiles running from the enlarged kitchen to the outdoor rear terrace, creating a strong connection between the house's interior and exterior.

The house is unified by a cohesive and colorful palette of finishes that when combined with an eclectic collection of furniture and art adds further layers to the renovations' overlay of traditional and modern sensibilities.

简净空间，艺术点睛

Neat and Crisp, Pin-pointing with Art

项目名称：宿雾峰
项目位置：美国加利福尼亚
设计公司：艺术线工作室
设计师：马克西姆
摄影师：蒂格兰

Project Name: Summit Circle
Location: California, USA
Design Company: Studio Artline Inc
Designer: Maxime Jacquet
Photographer: Tigran Tovmasyan

马克西姆设计的"宿雾峰"以建筑视野书写室内设计的精彩。897平方米的空间如同一个极简的绿洲。各色杂陈，如黑、炭灰、白、灰等。宁静的质感中，如同一个奢华的采石场，虽说平静，但却充满生机与能量。虽说以黑为基调，每一个色调、每一种质感共同捕捉着土、风、火、水等元素。每一堵墙都以定制的墙纸、石膏、油漆作为装饰。有的墙面呈以毛绒状，有的如皮质，有的如玻璃。当一切原本只能在梦中出现，耗时、耗资的空间自然成就了一处永恒：生态、祥和。

The Summit Circle home was designed by Maxime Jacquet, an architectural visionary specializing in interior sophistication. This 897 square meters house was designed to be a minimalistic oasis. The colors are almost all blacks, charcoal, white, grays, etc. It feels earthy soothing, and feels like you have escaped to a luxurious rock quarry filled with tranquil but lively energy. Despite the dark colors, each color and texture pop and create a vibrant mood that captures all the elements earth, wind, fire, and water. The most essential part to this design is that every single wall has been customized with a textural wall paper, plaster, or paint. Some walls appear to be tufted, some are textured fur walls, some are like grass, its everything you dream of. This is what made the home interiors so time consuming and expensive, but the impact it makes is timeless estate that is its own ecological world of serenity.

如果把硬装比作居室的身躯，软装则是其精髓和灵魂的所在。在室内装饰设计中，软装最能体现居者的生活态度与品位。软装，是结合风格、色彩、平面动线、生活方式所形成的一种氛围，是对不同文化与生活方式的尊重与理解。捕捉国际上各种时尚动态，不难发现2015年依然是以多元、个性的生活方式为主流趋势，将个性化元素与混搭、奢华、玩乐、田园等生活方式特点有机结合在一起，形成中式、欧式、现代、田园、地中海等多种软装风格共存的局面。风格的融合在体现出设计多元化的基础上，也从另一个方面强调了人们对不同生活层面的理解。

曾有名人言："最流行的风格就是没有风格"。在互联网时代之下，消费者可以更充分更自由地去享受家居的设计，让多元文化碰撞出火花，从而产生出各类"融合"的设计，即所谓的混搭风格，它糅合东西方美学精华元素，将古今文化内涵完美地结合于一体，充分利用空间形式与材料，创造出个性化的居住环境。混搭并不是简单地把各种风格的元素放在一起叠加堆砌，而是把它们主次有序地融合在一起，使之和谐并存。目前中西混搭是主流，其次还有现代与传统的混搭。在同一个空间里，不管是"中西合璧"，还是"传统与现代"，都要以一种风格为主，靠局部的点缀设计增添空间的层次。混搭体现了对多种风格的兼容，让不同的生活方式在同一环境下得以并存。可以预见，混搭以其开放性的手法和创意将会在未来持续很长的时间。

现代社会快节奏的生活方式给人带来了巨大的压力，充满玩乐模式甚至童趣的居住环境成为减压的方法之一。在软装设计中，要结合娱乐减压的诉求，灵活运用各种设计手法，让居住环境充满游戏感或娱乐感。因此具有不同的造型、充满童趣的色彩、新颖的材料、有趣味功能的设计，将会受到新一代的追捧。

中式风格——勾起怀旧思绪

中式风格是以宫廷建筑为代表的中国古典建筑的室内装饰设计艺术风格，气势恢宏、壮丽华贵、高空间、大进深、雕梁画栋、金碧辉煌，造型讲究对称，色彩讲究对比，装饰材料以木材为主，图案多龙、凤、龟、狮等，精雕细刻、瑰丽奇巧。但传统的中式风格装修造价较高，且具有太多的繁文缛节，显得沉闷，缺乏现代气息，因而很多人不太敢于挑战。

随着现代生活节奏的日益加快，现代风格大行其道。但有些人不满足于现代风格底蕴的苍白，想赋予其一定的文化内涵；部分接受传统中式风格的人也不满足其复杂繁琐和功能上的缺陷，想在保持韵味的情况下对其进行改变。于是，现代中式风格就产生了，渐渐且成为现代家装风格的主流。

新中式既不是对古典中式的简化，也不是用中式饰物装点简约之家，而是将现代和传统元素结合在一起，以现代人的审美对传统中式元素进行整合再处理，将复杂的传统秩序融入现代生活中，从而获得新的秩序感受，同时使用现代产品营造中式的意境，体现出优雅舒适的生活态度。色彩上，中国传统建筑和居室偏好黑、白、灰等素雅的颜色，因此新中式也以此为基调。家具只是略微保持了中式的影子，比如在细节的处理上借鉴传统形式，色彩则追随古人热爱自然天成，深沉的原木色比较流行；红、金等喜庆的颜色在饰品上运用得较多。软装更为花哨和热闹。

欧式风格——处处流露尊贵典雅

欧式风格主要是指西洋古典风格，这种风格强调以华丽的装饰、浓烈的色彩、精美的造型达到雍容华贵的装饰效果。受“重装饰，轻装修”的影响，古典风格抛却繁琐和严肃，在豪华大气的同时，更多的是惬意和浪漫。通过完美的典线，精益求精的细节处理，带给家人不尽的舒适享受，实际上和谐是古典欧式风格的最高境界。

地中海风格——享受碧海蓝天的清爽

文艺复兴前的西欧，家具艺术经过浩劫与长时期的萧条后，在9—11世纪又重新兴起，并形成自己独特的风格——地中海风格。地中海风格家具以其极具亲和力的田园风情、柔和色调，以及组合搭配上的大气很快被地中海以外的大区域人群所接受，并成为2015年家装流行趋势。

此风格具有独特的美学特点，一般选择自然的柔和色彩，在组合设计上注意空间搭配，充分利用每一寸空间，集装饰与应用于一体，在组合搭配上避免琐碎，显得大方、自然，散发出古老尊贵的田园气息和文化品位。

现代简约风格——行走在流行时尚前沿

随着时代的变迁、人们审美的变化，很多流行元素在变更，能经受住时间考验而经久不衰的简约风尚却一直被大众所追捧。日益加快的社会生活节奏，使人们越来越喜欢简洁而大方的现代设计风格。

现代风格是比较流行的一种风格，追求时尚与潮流，非常注重居室空间的布局与使用功能的完美结合，在追求时尚的同时讲究温馨舒适度。其装饰具有以下特点：由曲线和非对称线条构成，如花梗、花蕾、葡萄藤、昆虫翅膀以及自然界各种优美、波状的形体图案等，体现在墙面、栏杆、窗棂和家具等装饰上。线条有的柔美雅致，有的遒劲而富于节奏感，整个立体形式都与有条不紊的、有节奏的曲线融为一体。大量使用铁制构件，将玻璃、瓷砖等新工艺，以及铁艺制品、陶艺制品等综合运用于室内。

田园风格——沉醉在午后藤风花影

恬淡田园风格依然会在2015年持续热门下去，成为大家的最爱，尤其在家装领域，表现得更为淋漓尽致。纯美、田园、清新、淳朴的环境，让在钢筋水泥中穿行的人们唤起向往大自然生活的愿望。

恬淡田园风格的用料崇尚自然，在装饰上多以碎花、花卉图案为基础，给人浓郁的扑面而来的温暖温馨感觉，色调多是黄、粉、白等暖调。在织物质地的选择上多采用棉、麻等天然制品，其质感正好与乡村风格不饰雕琢的追求相契合，有时也在墙面挂一幅毛织壁挂，表现的主题多为乡村风景。最重要的是，田园风格的居室通过绿化把居住空间变为“绿色空间”，如结合家具陈设等布置绿化，或者做重点装饰与边角装饰，还可沿窗布置，使植物融于居室，创造出自然、简朴、高雅的氛围。

东方的意境，西方的时空

Oriental Inner, Western Existence

项目名称：上海万科五玠坊
软装设计：潘及
用材：榆木、皮、水晶、丝绒、银器、陶瓷、玉佩
面积：320 m^2

Project Name: Vanke (Shanghai) Wujie Fang
Upholstering Design: Eva
Materials: Elm, Leather, Crystal, Velvet, Silver Ware, Pottery, Jade Pendant
Area: 320 m^2

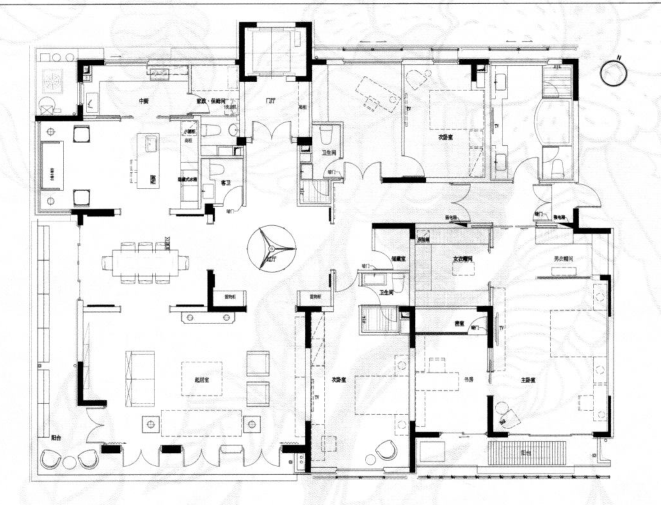

设计师常常在思考一个问题：什么样的设计风格，才是我们当代的主流？它既要保持我们文化的传承，又不能缺乏新时代的气息。

在本案的设计过程中，设计灌入了整个人物主题背景，并以其特点作为设计脉络。屋主夫妻拥有一儿一女，生活非常美满。男主人是金融投资者，留学归来，深受西方的教育；女主人则是一个热爱艺术的全职妈妈。他们有着相同的爱好——收藏画和摄影作品。女主人对艺术的品位和对生活的热爱在空间中表现得淋漓尽致。正是这样的人物背景，让本案空间弥漫着东西混合的特殊韵味。

设计师通过家具、面料、饰品等来表现东方元素，同时也利用了一些西方的方式和当代的表现手法，颜色的跳跃，材质的多元化，使得空间游走在东方文化和西方当代生活的碰撞中，呈现丰富且具有内涵的气质。

此外，设计还运用了 Hermes 的主题来延续他们的爱好，比如骑马等，价值不菲的 Hermes 马鞍、餐具、毯子等，为空间增色不少。来自意大利 Armanicasa、Flexform、Minotti 等世界最高端的家具品牌的家具，以及被认为是法国皇室御用级水晶品牌的 Baccarat 水晶灯，使空间更显显赫与尊

GOLDEN HORSE CUP

贵。设计甚至还运用了 Profoma Invoice 儿童家具的顶级品牌德国 Fink、AD，英国 Halo 的饰品，由知名设计师 Kelly Hoppen 操刀设计，更让空间靓丽精致。

There is a question designers tend to think over, what kind of design style can be the current mainstream to keep our cultural heritage still with atmosphere of the new era.

During the design, character theme is taken as the backdrop with its traits as the design vein. The host couple has one daughter and one son, happy in their life. The host, an investor of finance and returner from abroad is affected by the western education very much while the hostess, a full-time mother with a keen sense on art. Both share a hobby to collect pictures and pictures. The hostess's taste of art and love for life are revealed incisively and fully throughout the space. Such a context leads to the mashup of the east and the west.

Items of furnishings, fabric and accessories are used to express oriental elements, with western expressions and modern techniques. The diverse materials allows for a collision between the eastern culture and the western life to present the rich connotation.

Another theme of Hermes is employed to continue their common hobby, like horse riding. Of great value, Hermes saddle, dining ware and carpet boast the spatial glory. The highest-level furniture brand, like Armanicasa, Flexform and Minotti, as well as Baccarat chandelier, imperial consumer products in France compliments the spatial dignity. Halo accessories from Britain, and top name brand of Children furniture, like Fink and AD from Germany make the space more beautiful. And some are even done by Kelly Hoppen.

东方帝国 人文艺术触动美感奢华

Empire East: Cultural and Artistic Texture to Touch Luxury

项目名称：台湾东方帝国
项目公司：天坊室内计划
设 计师：张清平
面积：363 m^2

Project Name: Empire East
Design Company: Tianfun Interior
Designer: Zhang Qingping
Area: 363 m^2

空间是彰显屋主生活品位与态度的画布，依循着不同的空间故事，紧密掌握动线、造型、格局、材质与色调，使空间内的形、色、光、质等构成和谐的律动，因此环境与心境彼此相融呼应，塑造出独特的氛围渲染着感官知觉，而这是一个描绘人文、音乐与艺术邂逅的画作。

流畅的格局开展了空间的气度，引领着感官感受。空间的质感与细节，古典优雅的语汇以现代风格的手法演绎，运用金属、石材及玻璃完美地融合古典的当代风尚。丰富的石材纹理被利落线条收整，微妙的对比产生秩序美感，实现了视觉上的沉着稳定，却也营造出饱满而具有层次韵律，创造出不同角度的空间风景，映衬着屋主的品位收藏、知识累积的书籍及音乐的热情，不只流露奢华迷人的时尚表征，更多了一分人文与艺术淬炼的深度。

卧房中仍可以看到以音乐为灵感的设计巧思，有如乐曲般流畅的天花线条，或钢琴黑白琴键的衣柜设计，在华丽尊贵之外带来无限空间想象。宽阔的开

窗引入充足明朗的光线，光影成为连结内外空间最好的媒介，也为空间渲染出明暗。音乐艺术与空间艺术的交汇，让人体会每一刻美好生活的感动。

The interior space can actually be a canvas on which to bring out personal taste and attitude toward life. In different settings, lines, modeling, structure, material and hues are in control so items of shape, color, light and texture all together make a harmonious rhyme. And the surroundings and the mental state can echo with each other to set up a unique ambiance to render sensory perception. A real paining it is that runs into humanity, music and art.

A flowing pattern makes a good contribution to overspreading the spatial air, leading sense organ to feel the texture and detail. Classical vocabulary is interpreted with modern approaches while metal, stone and glass are fused into an impeccable state. The wealth of stone gain is softened with neat and crisp lines, a subtle contrast that generates an orderly aesthetics in realizing the visual clam and dynamic rhyme, out of which comes various perspectives to appreciate the space. The taste of personal collection, and the love for books and music are not only embodied on surface, but show more of the depth of culture and art. The bedroom continues the inspiration from music, like ceiling lines that extend like the flowing of music, and the piano-key wardrobe to stimulate limitless imagination besides magnificence and luxury. Large windows introduce more light, which with its shadow complete the media to link the internal and the external as well as render the bright and the dark for the space. The communication between music and art are friendly for people to appreciate the touch of good moments.

HELLO KITTY

雍容华贵巴洛克 洋洋洒洒荟中西

Elegant Baroque of the East and the West

项目名称：成都万科·五龙山别墅大独栋
开发商：万科地产
设计公司：深圳创域设计有限公司 / 殷艳明设计顾问有限公司
设计师：殷艳明、张书
用材：石材、墙纸、拼花木地板、木饰面、拼图马赛克
面积：660 m²

Project Name: Five-Dragon Villa, Vanke (Chengdu)
Developer: Vanke Real Estate
Design Company: Shenzhen Creative Space Decoration & Design Co., Ltd.
Designer: Ying Yanming, Zhang Shu
Materials: Marble, Wallpaper, Parquet Flooring, Veneer, Parquet Mosaic
Area: 660 m²

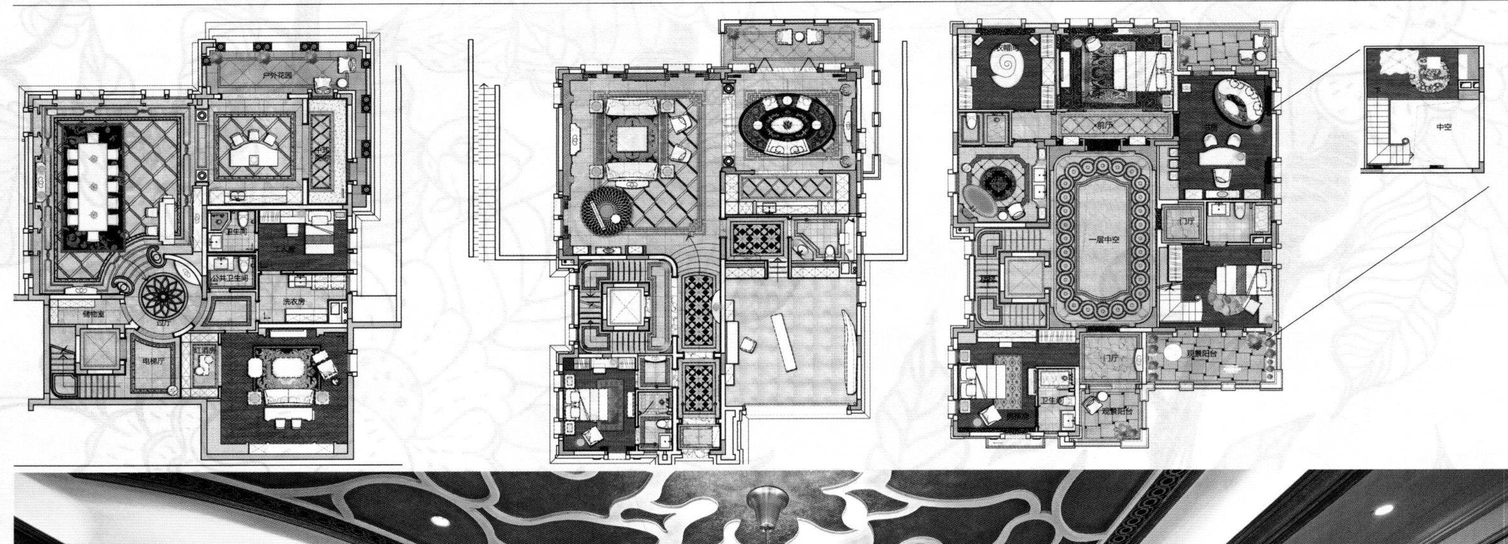

五龙山别墅大独栋位于成都北部新都区，别墅群背山面湖，是大隐于市的优雅栖居之地，圈层文化业态的形成缘于商圈和居住群体常年往来于世界各地，倾心于传统欧式生活方式与情调，本案以独特和具有创新精神的巴洛克风格打造一个新古典、宫廷式奢华风格的作品，符合项目注重家族传承的高端定位和设计诉求，融中于西的精神理念也贴近了客户群体对贵族化生活品质和文化氛围的需求层次。

本案强调软硬装一体化的设计理念，并通过空间与软装陈设的设计语言去解读巴洛克风格的特质与亮点，由空间解构开始，这是本案立意的一个高度，任何风格只有在适合的空间中绽放，才能彰显其精神价值与独特的个性，我们突破以往传统的别墅定位，以全套房型空间、并利用空间高度加入楼层之间的跃层空间双管齐下，从平面布局伊始便把巴洛克风格的奢华、恢宏与浪漫主义色彩的灵动赋予了空间强烈的立体感。从概念色彩体系、灯光体系，到艺术文化解读和造型体系几个方面，把十七世纪末巴洛克风格盛行时期的雍容华贵与同时期在中国传教的意大利画家郎世宁的宫廷绘画艺术相结合，巧妙而自然的中西合璧，给作品予以生动的依托。

位于一层的客厅会客区以宝石蓝色系为主，在西方宗教传统中宝石蓝是皇族与贵族钟爱的颜色，迷人而优雅，处处彰显贵族的气息。壁炉背后的黑金手绘壁纸，展现出主人对欧洲文化深厚底蕴的流连与玩味。整体设计方正大气，沙发群组与壁炉、吊灯及挂饰相映生辉，天花形态曲直相生，图案与光影交汇，展示了巴洛克风格动态中的平衡美感，古琴的设置让空间在精神层面上有了更高品位的追求。

女士餐茶区以酒红色为主，区域的设置突显了人性化的关怀和优雅生活方式的尊贵，孔雀蓝与羽毛让闲雅的空间有了鲜活的气息。

二层是主卧，空间内的所有家具与软装配饰格调相同，地面古典图案的咖啡色地毯与拼花木地板、金色雕花的屏风，柔化了居室的硬朗质感，在整体营造奢华氛围的同时，床头绢上绘画《百骏图》更突显低调中的奢华。

浴室的设计紧致、优雅。孔雀摆件姿态怡然，绚丽夺目。洗手台采用烟玉大理石，搭配古典贝壳马赛克的铺砌，定制的金色椭圆镜点缀，营造出一个浪漫、舒适的空间氛围，这不是一个传统意义上的浴室，而是可以品酒、身心愉悦的休憩场所。

中庭空间承上启下，水晶吊灯与绽放的花形、地面流动的圆形图案都在优雅中传递空间的气质与精彩，形与意、态与势展露出瑰丽奢华的贵族风范。

负一层大厅改设为宴会厅是具有仪式感的一个重要空间，宴会厅空间布局与陈设热烈、激情而又华丽，巴洛克激情艺术的气氛展露无遗，金色与蓝色的碰撞给人强烈的视觉震撼。

男孩房以海蓝色为主，同时在局部增添了白色，运动、时尚，色彩沉静，高贵中渗透出年轻人的气质与修养，给人带来创意、时尚的动感。

女孩房粉紫色的主色调让人觉得甜美温馨，跃层空间的儿童活动区开放而有浓厚的童话气

息，缤纷的色彩洋洋洒洒造就了一片童心世界的意境。
父母房色调沉着、层次丰富，中式纹样图案与配饰青花瓷相得益彰，在光影中传递出美好的感性体验。父母房色调沉着、层次丰富，中式纹样图案与配饰青花瓷相得益彰，在光影中传递出美好的感性体验。
多功能厅空间色彩浓烈，打破理性的宁静和谐，强调激情的艺术，家具、灯具、饰品、布艺……它们和着空间的韵律节奏、功能主题的转换，具有浓郁的浪漫主义色彩。
本案立意非新，贵在完整贯彻了软硬装一体的设计初衷，细细研磨，终有回响；准确的定位与理念是设计落地的根本，一个成熟的设计作品必是贴近生活融入历史的，真正打动人的不是繁复与冗繁的造型、线条，也不是流金溢彩的水晶灯饰与装饰，而是对设计的追求与执着。沉醉奢华而又雍容尔雅，华丽只是表象，空间的情感才能真挚动人。

A project this space makes a real reclusive land in a metropolis setting with hills and lakes as its backdrop. People in business circle, regular travelers around the world with preference for European style, have now been taken as its potential consumers. The interior is thereby aimed for a luxurious creation of neo-classical, imperial palace, at one with high-end position of a family with its design both eastern and western, more considerate to meet the aristocratic demands of life quality and cultural atmosphere.
With equal emphasis on integrating the decoration and the upholstering, the project turns to design vocabulary of space, furnishings and accessories to make interpretation of traits and light spot of Baroque. The process starts from spatial deconstruction, a high level the design for this project is destined for, for as long as it can fit in well with a space, a style can be brought out its spiritual value and unique personality. Villa stereotype is broken away by making the best use of the holistic

area and the height with split-level design employed. Even from the plan layout, the luxury, the magnificence and the romantics exclusive to Baroque style altogether allows for a strong 3D sense. The popularity of grace and elegance of Baroque in the 17th century is combined with place painting of Giuseppe Castiglione, in terms of hues, lighting and art to offer the space something vivid and life-like.

The living room on the 1st floor is dominated with sapphire, a color more liked by imperial family in western religion for its grace to embody the noble sense everywhere. The hand-painted black gold wallpaper behind the fireplace reflects personal passion and enthusiasm for European culture. Here is of grandeur and generosity, where the sofa group, the fireplace, the chandelier the pendant, and the ceiling are set off each other, with lines and curves, patterns and light and shadow indicating a balanced aesthetics of the dynamic Baroque. Meanwhile, the seven-stringed musical instrument reaches the spiritual pursuit to a new level.

Female dining area is mainly coated in red wine, where to highlight the humanized care and the graceful life. With peacock blue and feather,

here becomes more fresh and alive.

The 2nd floor is for the master bedroom, whose furnishings remain the same with the upholstering in style: the coffee of classical carpet, the parquet wood flooring, and the screen of gold carving soften the stiffness, and the silk painting of *All Pretty Horses* above the bedhead furthers the reserved luxury in setting off the whole sumptuous air.

The bathroom is compact but elegant. The peacock looks easy, brilliant and splendid. The washbasin is of Misty marble and classical shell mosaic. The custom gold oval mirror builds up the spatial romance and comfort. It's more a resting place where to drink and relax than a traditional bathroom.

The atrium make a linking connection for the upward and the downward, where the chandelier, the floral model, and the dynamic circle on the ground convey manners only nobles have from perspective of physical and spiritual level, and posture and power. That is magnificent and luxurious.

The basement is refurbished into a feast hall with more ritual sense. Its furnishings and accessories are of passion and extravagance to incisively and fully the Baroque art with the collision of gold and blue to exert a strong visual impact.

The boy's room is of ocean blue with white available now and then. Out of the sports, the fashion, the staidness and the noble, comes the young nature and cultivation to bring creative and modern innervation.

The persistent pink in the daughter's room is sweet and warm, where the spilt-level area is open and of strong sense of fairy tale, and colors accomplishes an artistic concept belonging to a child's heart.

The room for the elders is for

sure staid and steady in terms of hues. Against the abundance of spatial hierarchy, Chinese pattern and accessories of China blue porcelain are brought out the best of each other, conveying a good emotional experience with light and shadow dancing.

The multi-function hall is colorful to break the rational peace and harmony, accentuating the passionate art, furniture, light fixture, accessories and fabric art, all echoing to the tone of the space with theme shifted from one to another to

overspread a strong sense of romantics. This is a design not deliberately met for innovation, where to carry out the integration of decorating and upholstering to make a proof that, an accurate position and concept is the fundamental of a design, a mature project should be fused into life, and what to move people is the pursuit of design instead of the complicated or complex modeling, lines, chandelier or design. Even sumptuous, genteel and magnificent, all can be artificial, because spatial emotion in its real sense can be sincere and cordial.

一石一木一份心意，重新写就家的故事

Home of Stone and Wood

项目名称：农场别墅
设计公司：洛佩斯建筑设计
设计师：洛佩斯
摄影师：杰马

Project Name: Dores Do Indaia Farm
Design Company: Gislene Lopes Arquitetura e Design de Interiores
Designer: Gislene Lopes
Photographer: Jomar Braganca

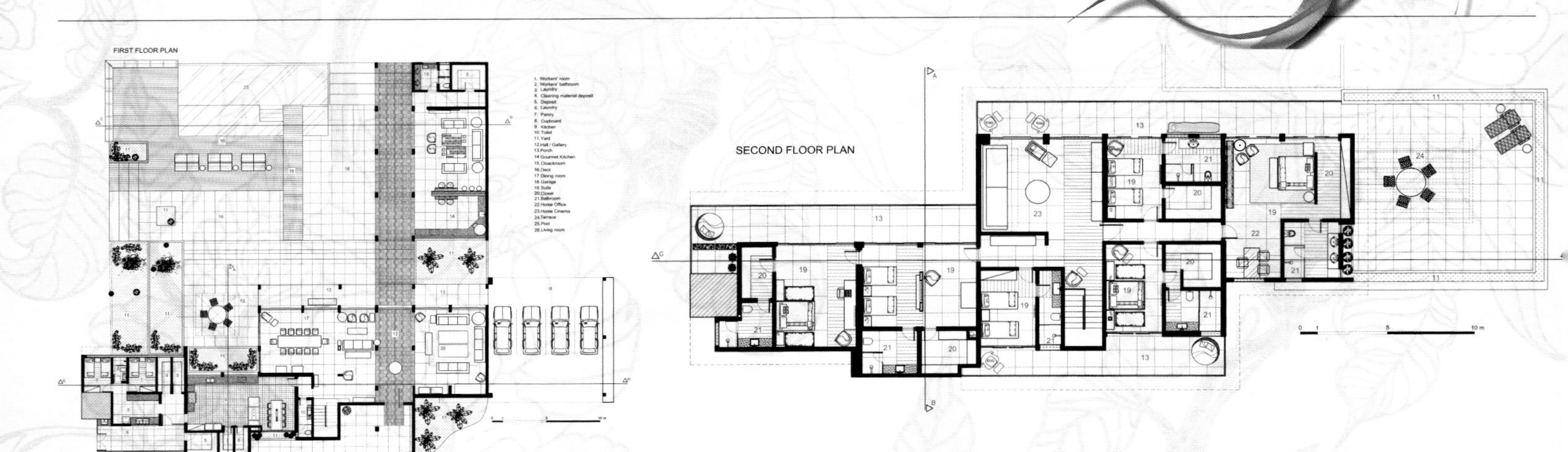

家庭农场的设计不但要让空间充满故事，有内涵，更不能失去建筑的现代感。设计借助于屋顶大大的挑檐及富有历史沉淀的设计，赋予空间清爽线条的同时，又保持了设计语言上的现代感。一楼与服务区、餐厅、门廊、客厅相互隔离，中间以廊道相连。户外设有美食空间，并有泳池。二楼刻意把房间与室外阳台连为一体，充分地利用了外面的美景。
住户生活空间得到满足的同时，更得到了生活上的舒适、内心的舒畅。元素虽然不同，但无不体现尖端与美学，开阔的空间变得更为丰富。新式的用材集历史感、现代作品于一体，如水泥瓦、水泥地板、回收木等。农场建筑基地广阔。别墅战略性地位于入口附近，除了拥有花园美景，通往建筑的路旁还有几个反光水池。
新式的建筑设计空间里，如何保持家族的元素，并把这些元素与新式建材融为一体是本案设计的难点。清爽、优雅的设计语言，选用与建筑线条保持一致的用材，并参照巴西米纳斯吉拉斯州农场的做法，最终促成了本案优雅现代的设计。
通过本案，我们不但看到了优雅建筑美丽的一面，更看到了清爽、现代的线条。

As it is a family farm, the idea was to create a work that would bring the story of the place, without losing the contemporaneity of the construction. The architecture combined with the clean style of the farms from Minas Gerais expressed through the roof with large overhangs and decoration with strong historical references brought great elegance to the project, without losing modernity in its language. The lower floor is separated for the service area, dining rooms, porches and living room, connected by a covered walkway with the gourmet outdoor area and pool. The distribution of the second floor rooms had the intention to connect the rooms to the outside balconies, taking advantage of the beautiful view.
All the rooms were thought to offer their residents their

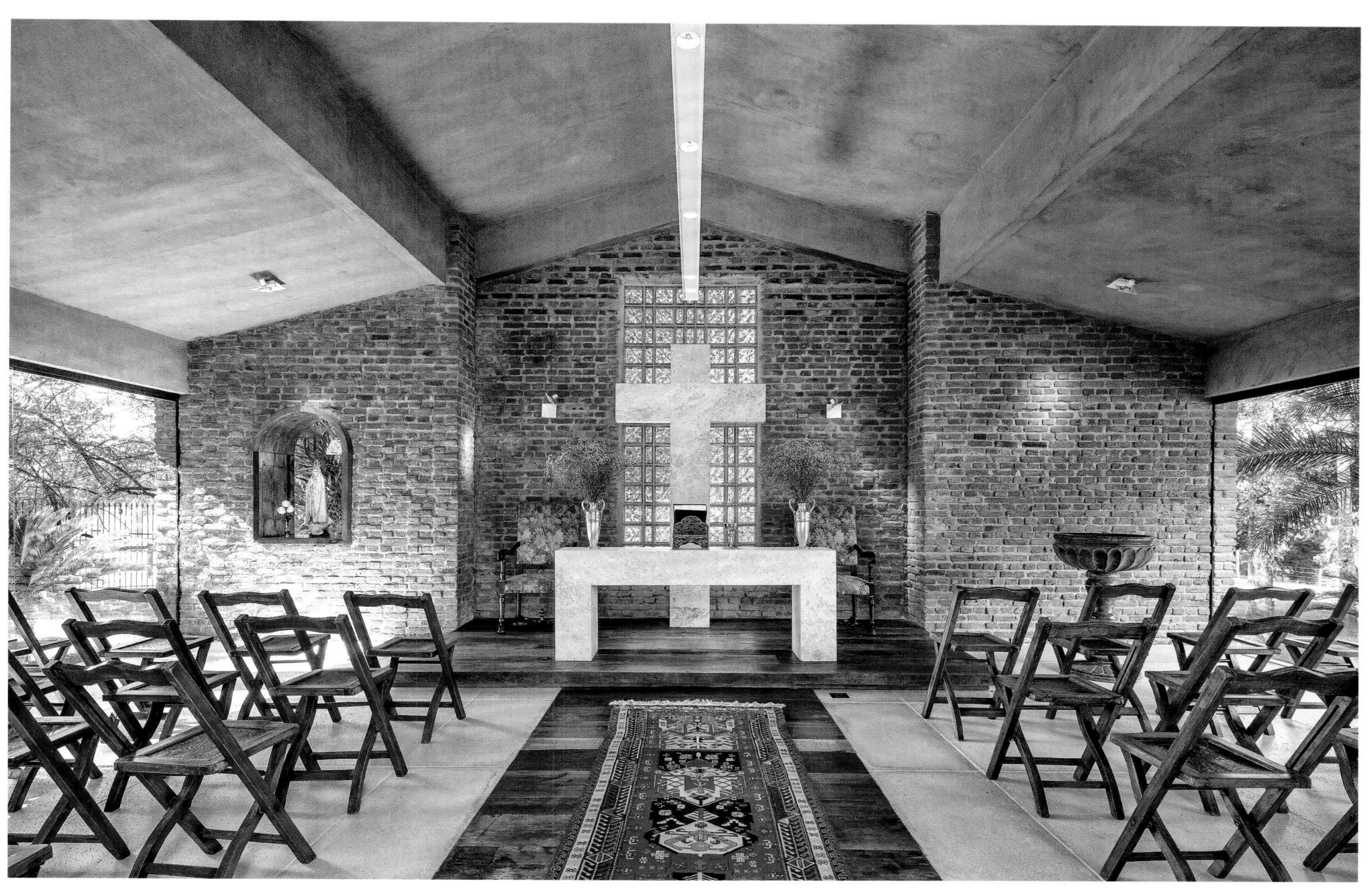

personalities translated in space and their needs met in a comfortable way. The objective was to enrich the wide spaces and bring different elements with sophistication and authenticity. The choice of new materials follows the line of the project to combine historical references with a contemporary work, such as using hydraulic tiles, cement floors, demolition wood and others.

The farm has a very wide building sit. The implantation of the house is strategically positioned near the entrance with space for a beautiful garden and reflecting pools on the way to the house.

There was big concern about preserving the family elements in the context of the new house and decoration of the rooms, enriching them when using them with the new materials. Searching for a clean and elegant language, the choice of the new materials follows the same line of the project, to ally the historical references of the farms from Minas Gerais with a contemporary and elegant project.

The final result makes us admire the beauty of an elegant architecture with clean and contemporary lines.

JARDINS BOTÂNICOS DO BRASIL

收藏者之家

Apartment of a Collector

设计公司：基里尔室内设计
设计师：基里尔

Design Company: Kirill Istomin Interior Design & Decoration
Designer: Kirill Istomin

本案与其说是公寓，倒不如说是一个典藏的场所。长长的廊道，与众不同的繁杂背景，数不清的壁龛都构成了设计的难题。业主可谓是收藏的集大成者，有图案、绘画、摄影等。收藏年代横跨古今。中性的墙面，特别的灯光，真正地让本案成了一个博物馆般的艺术空间。基里尔的出色之处就在于并没有让本案成为一个仅供人参观的空间。相反，设计着手时，依然以家居为本，所有的艺品只不过是用来装饰的细节。图案多出现在廊道里，而绘画作品则出现在客厅。经过与业主沟通，对一些空间进行了改变，而有的艺品刻意没有出现在空间里。艺术品主题的两个房间为主卧、女儿房。卧室里，巨幅绘画作品置于镜面壁龛里。女儿房里，修长、带有郁金香的作品用作了房间的装饰板。旧时装饰留下的18世纪的家具在本次设计里予以保留。不仅适合着现代的内部装饰风格，也彰显着苏联时期的现代艺术。

This spacious apartment for a collector, his huge collection and small family is made out of two smaller apartments. That's why space was rather difficult to decorate – it had long corridor, lots of room with unsual settings, niches. The collection of the client was also rather unusual – it is the mixture of graphics, paintings and photography. It has both old and modern art. It would have been easier to create kind of a museum – with neutral walls and special lighting, so that the pictures look at their best. But Kirill chose another way – his idea was to make an apartment for

›eople, not for pieces
›f art. So all the art
›bjects were used as
lecoration details. The
oncept was that all the
graphic is represented
n the corridor and
›aintings – in the
iving room. But while
vorking on the project
stomin together with
he client changed
›laces and even left
ut some of the pieces.
'he only two rooms
hat were determined
›y art pieces were
naster bedroom and
aughter's bedroom.
n the first case it was
lear that the huge
ainting will be placed
n a mirror niche and
vill dominate the
oom, so there was
o use to resist that.
n the second case,
n the daughter's
edroom, the narrow
all paintings with
ulips are used as
ecorative panels.
'he client wanted
o use some of the
ntique 18th century
urniture pieces from
ne previous interior.
omehow they
erfectly matched both
nodern interior and
nodern Soviet art.

在黑与白的基调上 描述历史与现实交织的梦境

The History and the Modern Interlaced on Black and White

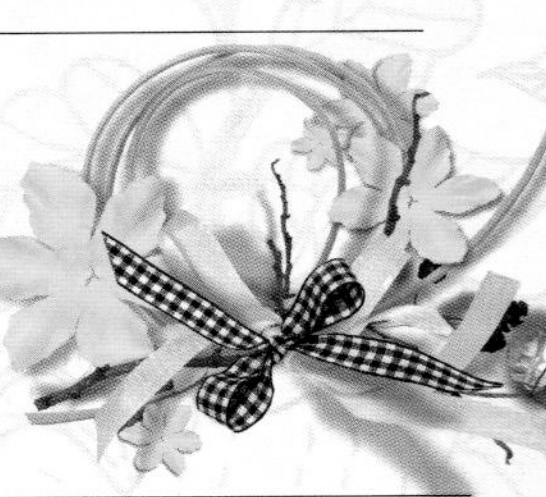

项目名称：索菲亚别墅
设计公司：梦设计
设计师：埃琳娜
摄影师：伊戈尔
面积：1 150 m²

Project Name: Villa Sofia
Design Company: Dream Design
Designer: Elena Dobrovolska
Photographer: Igor Karpenko
Area: 1,150 m²

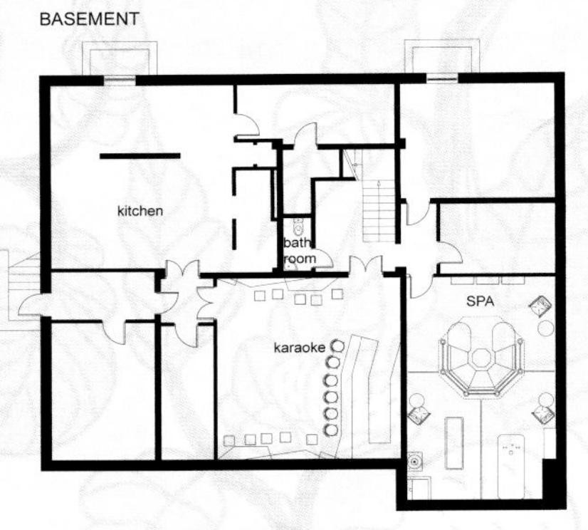

“索菲亚别墅”是个概念化设计，其实是个酒店。所在量体极富历史底蕴，由19世纪知名建筑师尼古拉·克拉斯诺夫设计。
如今的酒店，是内外相互和谐的统一体。不同风格在此跨越了各个时代。每个房间各自的魅力源自于精心考究的细节，色度及铺陈而得以彰显。典型的埃琳娜设计师之手笔。法式的优雅经典与东方的魔幻相互辉映。东西方于其合为一体，是本案的、也是世界的，堪为典范。很多建筑师在此都可以找到属于自己的独特灵感。

装饰所用材料与众不同：自然的山石、木料。房间家具摆设多为意大利设计师为空间量身定做。所有的窗户全为木质框架。每扇窗户如同框景一般，透视着克里米亚半岛的秀美山水，为室内引进徐徐的清新空气。拱形窗，伴着优雅的经典家具书写着现代的舒适。等离子体显示器、迷你酒吧、保险柜、奢华的窗帘自是不必陈述。就连独立的卫生间，也有着优雅的设计，以最高的现代标准配备。
12间客房，7种不同的优雅的风格。无论是经典法式风，还是东方风都以摩尔式的建筑元素作为衬托。两种风格最受客人欢迎。

Villa Sofia is a place with idea. The hotel was opened in a historical building, constructed by Nikolay Krasnov, a prominent architect of the XIX century.
A harmonious relationship of the exterior and the interior creates the wholesome character of the hotel. In Villa Sofia you can feel the atmosphere of unique combination of different styles and times. Each of the rooms has its own charisma, which is emphasized by elaborate details,

2 FLOOR
bedroom 5
bedroom 6
bedroom 1
bath room
bath room
bathroom
bathroom
bathroom
bathroom
bedroom 4
bedroom 3
bedroom 2
3 FLOOR
bedroom 10
bedroom 11
bedroom 7
bath room
bath room
bathroom
bathroom
bathroom
bedroom 9
bedroom 8
living room 7
4 FLOOR
bedroom 12
bath room
living room 12

colors and accessories. This is typical for all the objects of Elena Dobrovolskaya. You can experience the refined classic French charm or feel the magic breath of the East. While the combination of East and West is a favorite source of inspiration for many architects, Villa Sofia combines the best of both worlds and it is not a cliche.

The business card of this hotel is the exclusive use of natural materials in the decoration: natural stone, wood, marble. Most of the furniture in the rooms was created specifically for Villa Sofia by Italian designers according to individual projects. All of the hotel's spaces have wooden window frames, and each of the windows looks out to the gorgeous Crimean landscape and fills your lungs with fresh sea air. Arched windows and refined classic furniture coexist with modern comfort and amenities – such as plasma displays, mini-bars and room safes. Each room has luxury curtains and linens and a separate bathroom with a unique design, equipped to the highest modern standards.

The twelve guest rooms of the hotel are decorated in several refined styles. The classic French design and the Oriental, with Moorish style elements, are the most popular among the visitors. In the French style rooms you have a fresh palette, wallpapers with rich floral patterns and lightweight furniture, and in the Oriental interiors – blue purple maroon colors and lighting design typical of hot countries. The use of contrast colors in modern European interiors is softened by textiles and ornate decoration. Each room is different, yet, what remains unchanged is the use of natural materials and the care for environment.

图书在版编目（CIP）数据

国际软装设计流行趋势 / 黄滢，马勇 主编 . – 武汉：华中科技大学出版社，2015.7

ISBN 978-7-5680-1023-8

Ⅰ . ①国… Ⅱ . ①黄… ②马… Ⅲ . ①室内装饰设计 – 世界 – 图集 Ⅳ . ① TU238-64

中国版本图书馆 CIP 数据核字（2015）第 157651 号

国际软装设计流行趋势 黄滢 马勇 主编

出版发行：华中科技大学出版社（中国 · 武汉）
地　　址：武汉市武昌珞喻路 1037 号（邮编：430074）
出 版 人：阮海洪

责任编辑：熊纯　　责任监印：张贵君
责任校对：岑千秀　　装帧设计：筑美文化

印　　刷：中华商务联合印刷（广东）有限公司
开　　本：965 mm × 1270 mm　1/16
印　　张：21
字　　数：168 千字
版　　次：2015 年 9 月第 1 版 第 1 次印刷
定　　价：338.00 元（USD 67.99）

投稿热线：（020）36218949　　duanyy@hustp.com
本书若有印装质量问题，请向出版社营销中心调换
全国免费服务热线：400-6679-118 竭诚为您服务